Lena Zentner, Sebastian Linß
Compliant systems

Also of interest

Drilling Technology
J.P. Davim (Ed.), 2018
ISBN 978-3-11-047863-1, e-ISBN 978-3-11-048120-4,
e-ISBN (EPUB) 978-3-11-047871-6

Topological Optimization of Buckling
B. Bian, 2018
ISBN 978-3-11-046116-9, e-ISBN 978-3-11-046227-2,
e-ISBN (EPUB) 978-3-11-046127-5

Progress in Green Tribology
J.P. Davim (Ed.), 2017
ISBN 978-3-11-037272-4, e-ISBN 978-3-11-036705-8,
e-ISBN (EPUB) 978-3-11-039252-4

Metal Cutting
J. P. Davim (Ed.), 2016
ISBN 978-3-11-044942-6, e-ISBN 978-3-11-045174-0,
e-ISBN (EPUB) 978-3-11-044947-1

Lena Zentner, Sebastian Linß

Compliant systems

Mechanics of elastically deformable mechanisms, actuators and sensors

DE GRUYTER

Authors
Univ.-Prof. Dr.-Ing. habil. Lena Zentner
Technische Universität Ilmenau
Department of Mechanical Engineering
Compliant Systems Group
Max-Planck-Ring 12
98693 Ilmenau
lena.zentner@tu-ilmenau.de

Dr.-Ing. Sebastian Linß
Technische Universität Ilmenau
Department of Mechanical Engineering
Compliant Systems Group
Max-Planck-Ring 12
98693 Ilmenau
sebastian.linss@tu-ilmenau.de

ISBN 978-3-11-047731-3
e-ISBN (PDF) 978-3-11-047974-4
e-ISBN (EPUB) 978-3-11-047740-5

Library of Congress Control Number: 2019931431

Bibliographic information published by the Deutsche Nationalbibliothek
The Deutsche Nationalbibliothek lists this publication in the Deutsche Nationalbibliografie; detailed bibliographic data are available on the Internet at http://dnb.dnb.de.

Cover image: Feng Yu/iStock/thinkstock
Printing and binding: CPI books GmbH, Leck

www.degruyter.com

Preface

Compliant systems are characterized by complex deformation and motion behavior, which complicates both their modeling and ultimately their application. This book presents methods both of classifying and of modeling compliant mechanisms, actuators, and sensors according to diverse criteria, enabling or simplifying their selection, modeling and design. The content of this book is based on many years of experience gathered at the Compliant Systems Group at Technische Universität Ilmenau, and is the subject of a number of classes offered to students of the Department of Mechanical Engineering.

This book is intended both for students of engineering, as well as those who have already qualified. Further, it is written for anyone interested in compliant systems, or who wishes to implement them. Even without special scientific education, the second chapter of this book, which describes the classification of compliant systems, will provide an understandable overview of compliant mechanisms, actuators and sensors, as well as possibilities for their application. The third chapter, regarding the modeling of compliant systems as rigid-body systems, is based on the linear theory of material strength science. The content of the fourth chapter, regarding the modeling of large deflections in compliant mechanisms and actuators, offers an introduction to non-linear theory and includes detailed derivations, while ensuring that the calculations and results remain intelligible. This is due in part to mathematical transformations that are purposefully designed to be uncomplicated. The examples given in the fifth chapter have been taken from a variety of applications, and their results, often taking the form of mathematical relationships between different model parameters, can be reused for similar cases. The final chapter suggests methods of synthesizing compliant mechanisms with concentrated compliance, as well as for the design of flexure hinges with special consideration to optimized notch contours.

Chapters 3 to 5 and partially Chapter 2 of this book are translations of the German book "Nachgiebige Mechanismen" ("Compliant Mechanisms") by Lena Zentner [114] with several improvements and extensions. Chapter 6 and partially Chapter 2 contain knowledge based on several research papers by Sebastian Linß [58]-[66].

Thanks are due to all members of the Compliant Systems Group at Technische Universität Ilmenau, who have helped increase the competence and experiences of the field of compliant mechanisms, actuators and sensors through their own efforts, the supervisions of student theses and joint discussion at the Group. Special thanks go to Prof. Valter Böhm, his cooperation was fundamental to the classification of compliant systems. Last but not least, thanks go to Matthew Partridge for supporting the English translation of this book.

Ilmenau, January 2019 Lena Zentner and Sebastian Linß

https://doi.org/10.1515/9783110479744-005

Contents

1 Introduction

Compliance is no longer considered a disadvantage of mechanical systems. Instead, the resulting advantages, such as the utilization of elastic restoring forces, possibilities for energy storage, structure coherence etc., are used specifically to provide a system with new qualitative properties. In many application areas, such as in medical technology and human-machine interaction, specific and differentiated compliance is even a requirement for technical systems ([7], [125]). Conventional rigid-body mechanisms are being successfully supplemented or replaced more and more with compliant systems even in classical branches of mechanical engineering, often to take over motion or force transmission tasks, for example in gripping devices (Fig. 1.1). This tendency is being facilitated by the development of new materials and corresponding production technologies. Highly-elastic materials allow for actuators that can be designed and integrated into a compliant system in such a way that such systems take on inherent actuator properties. In combination with the use of functional materials, a system can be given inherent sensing properties, while allowing for a compact design and higher multi-functionality of the system as a whole.

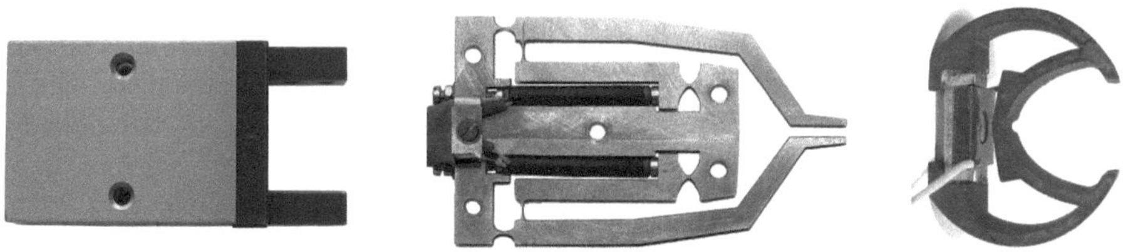

a b c

Fig. 1.1: Examples of different gripper systems; a – conventional two-finger parallel gripper; b – compliant parallel gripper with concentrated compliance and piezo actuators; c – compliant parallel gripper with distributed compliance and inherent actuation based on ferroelastomers

By first investigating thoroughly its complex behavior, both in deformation and motion, a focused application of the actuators and functional materials can be guaranteed, while also ensuring a sensitive design for the compliant system as a whole. The most suitable method is by using model-based investigations, especially if an analytical model can be constructed to this end. The complex relationships between different parameters can thus be written in a transparent way, and dependencies between them can be revealed. Even if a solution cannot be achieved entirely through analysis, the resulting relationships between the mechanical parameters of a system can help in understanding it and comprehending its behavior. Compliant systems are considered under the effects of different loads, which are mostly caused by moments and forces or distributed moments and forces designated as line loads or area loads. These loads as well as other vector parameters, such as radius vectors or displacements are characterized by both a direction and a value. The terms for the vectors are given in bold text, e.g. **F** for a force vector, while the values of the vectors or scalar parameters are given

https://doi.org/10.1515/9783110479744-011

in italics, such as the value of a force: *F*. In order to differentiate them from vectors, matrices are written in bold and underlined: $\underline{\mathbf{T}}$.

In the following chapters, classifications of compliant systems are first presented, which have been collected and systematized in the field of compliant mechanisms and actuators. Compliant systems with linear elastic properties are the focus of the modeling. Linear theory, which leads to the linear differential equations, is used to describe the actual behaviors of elastic deflections in a system undergoing small deflections. For large deflections, non-linear theory should be used when constructing a usable model for given deflections. Both methods of modeling are based on classical methods and have been both expanded upon and methodically generalized, in order to account for special cases, while also providing a formalism for simplifying the modeling process.

Before choosing an analytical method for describing the deformation of a compliant system, the character of this system, mirrored in its geometry and material, should first be established. In this step, a decision is made about which parts of the system must be modeled as compliant bodies and which as rigid bodies, thus excluding the compliant system parts to be modeled from the mechanical system as a whole. Subsequently, a decision must be made about which modeling methods are suitable for a given situation, based on the behavior and the uses of the system. The behavior of a system is understood here as both the deformation behavior of a compliant system part, as well as its motion behavior. The expected deformations, regarded as either large or small deflections, should be modeled accordingly using either a non-linear or linear theory. The intended application area of the system and corresponding objectives should also be taken into account. The next important step before formulating a mathematical model is determining the boundary conditions. Decisions are made here about where in the system loads will be applied, and which loads can be modeled as idealized concentrated forces and moments, or as distributed forces and moments. This also includes decisions about the bearing of the compliant system parts, which determines the boundary conditions for forces and moments.

After completing the modeling process, the calculated results of the model and the model-based simulation should be evaluated, based on the assumptions made and the starting requirements. If, for example, a compliant element with a cross-section measuring a third of its length is described using the theory for thin rods, it can be assumed that the results will correlate less accurately, if at all, than real deformations in an actual thin rod. The success of a model-based investigation depends on a keen examination of modeling method theory and a critical consideration of the results.

Based on existing research, a simple, usable method, based on the rigid-body replacement approach, of synthesizing compliant mechanisms with concentrated compliance is presented in the final chapter. The numerous possibilities of realizing a flexure hinge are first listed and evaluated based on their uses in precision engineering technology. A synthesis method based on the rotation angles of all hinges is especially suitable for mechanisms with different optimized flexure hinge contours, and is explained using the example of a four-bar mechanism.

The derivations of the modeling methods and their solutions to the examples presented are kept both clear and concise, in order to minimize the time and effort required to become acquainted with the material and to learn the methods. The calculations of mathematical equations are carried out using the software Mathematica® and MATLAB®. Finite elements method (FEM) simulations are performed with ANSYS Workbench®. Where possible, analytical solutions have been preferred over numerical solutions.

2 Definition and classification of compliant systems

To give an overview of the possible applications of compliant systems, these must first be classified according to their characteristics. The structural considerations of compliant systems and their properties are especially important when planning and building accurate models. In this way, it is possible to consider only the characteristics of a system that partly or wholly influence its functionality. This simplifies the development and modeling of compliant systems for specific purposes. A *compliant system* can be considered as a compliant mechanism, a compliant actuator and/or a compliant sensor according to its function (Fig. 2.1).

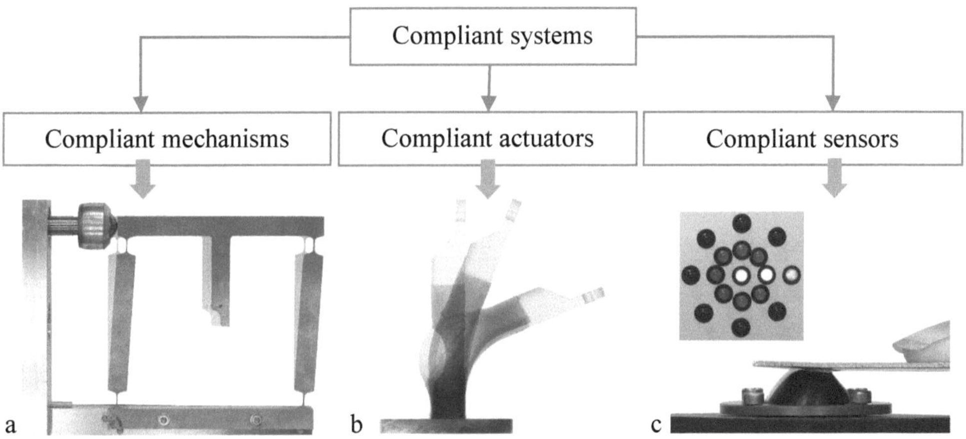

Fig. 2.1: Compliant systems; a – four-bar compliant path-generating mechanism with flexure hinges; b – compliant fluid-mechanical actuator made from a highly elastic polymer; c – compliant sensor used to detect shear forces

Compliance, as an important property of compliant systems, as well as its causes and effects on a system will be considered first of all. Compliant systems will accordingly be broken down according to their motion behavior. The definitions that follow in this chapter are based on the terminology for compliant mechanisms according to the International Federation for the Promotion of Mechanism and Machine Science (IFToMM), but also go further to supplement these. The following descriptions and classifications should assist in the selection, modeling and design of compliant systems, and are also used in later chapters.

https://doi.org/10.1515/9783110479744-013

2.1 Compliance

Compliance, as basic property of compliant systems should first be considered more specifically, thus allowing for a sensible and purposeful application in mechanical systems. Understanding the factors that can change or manipulate compliance, allows this property to be adjusted deliberately to external conditions. Thus, compliant systems can be designed in such a way that they can change their mechanical properties depending on variable conditions, using either a control system or of their own accord.

2.1.1 Classification of compliance

Compliance is the measure of the ability of a system, body or body segment to exhibit a deformation due to the action of external forces (reciprocal of stiffness). The term compliance and not elasticity is purposefully used here, because compliance includes the properties of elasticity, plasticity and viscoelasticity, in general, whereas elastically deformable components of compliant systems are the focus of this book. The compliance of a system element is determined by the value of displacement of a point due to an applied force.

Compliance can be separated into concentrated and distributed compliance, according to its geometric distribution within a body or a system ([116], [117]). A power of 10^1 can be taken as a comparative value. When the maximal measurement of a compliant segment is ten or more times smaller than the maximal characteristic length in the direction of the distribution of a compliant area ($L/l \geq 10^1$), this is regarded as *concentrated compliance.* Alternatively, in cases where a deformable area exhibits a distribution that is comparable to the maximum characteristic length of the system ($L/l < 10^1$), this is named *distributed compliance* (Fig. 2.2). Either the length of a system component or the entire component can be considered the characteristic length, depending on the theoretical view (compare Section 3.1 and Section 6.1).

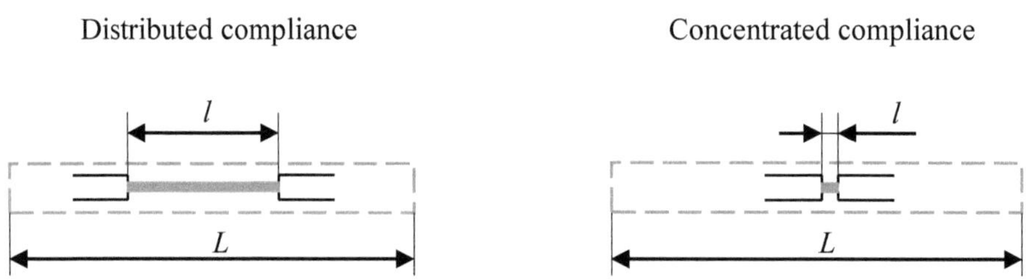

Fig. 2.2: Schematic depiction of the difference between distributed and concentrated compliance

The benchmark of 10^1 given above should be seen as an approximate value. This conception of compliance based on its geometrical distribution helps in the choice and application of modeling methods for compliant systems. If a compliant area is seen as a *coherent joint*, a difference can be drawn between a *joint with distributed compliance* and a *joint with concentrated compliance.* In Fig. 2.3, two examples of compliant mechanisms with coherent joints are given, that exhibit distributed and concentrated compliance.

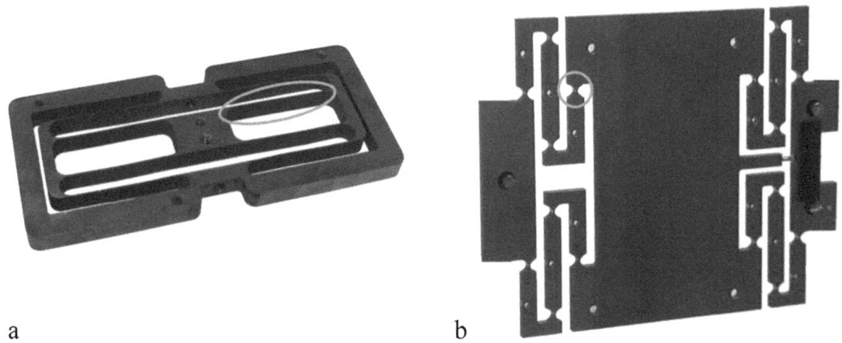

Fig. 2.3: Examples of compliant mechanisms with coherent joints; a – with distributed compliance; b – with concentrated compliance

2.1.2 Variation of compliance

Varying compliance has certain advantages, for example, the ability to adapt to a particular situation or a variable environment. Compliance exhibited by a system can be classified based on its capability to vary (Fig. 2.4). Many technical rigid-body and compliant systems show *constant compliance*. If the compliance is not constant and can vary due to variations of material or geometric properties, then this is considered as *variable compliance*. Variable compliance that re-establishes its original state after disappearance of the cause of its variation is named *reversibly variable compliance*. Conversely, *irreversibly variable compliance* is defined as a compliance where there is no return to the original state after the cause of the variation is removed. Such systems can be broken down initially according to their variations as follows.

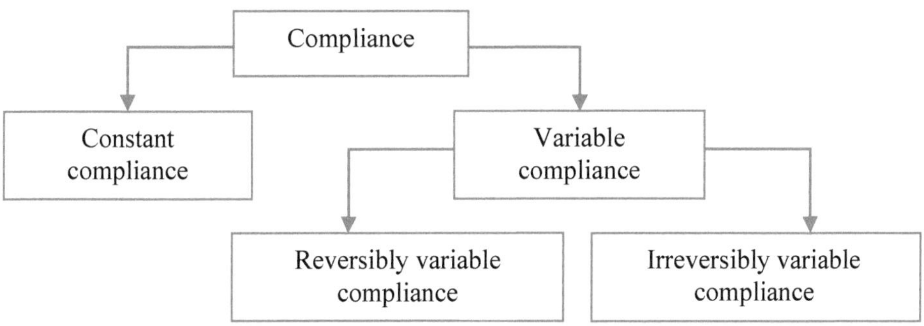

Fig. 2.4: Classification of compliance according to the ability to vary

Changing the compliance of an individual segment of a body can be achieved by influencing its geometry, that is, precisely designing it to this end, or through variable material properties (Fig. 2.5).

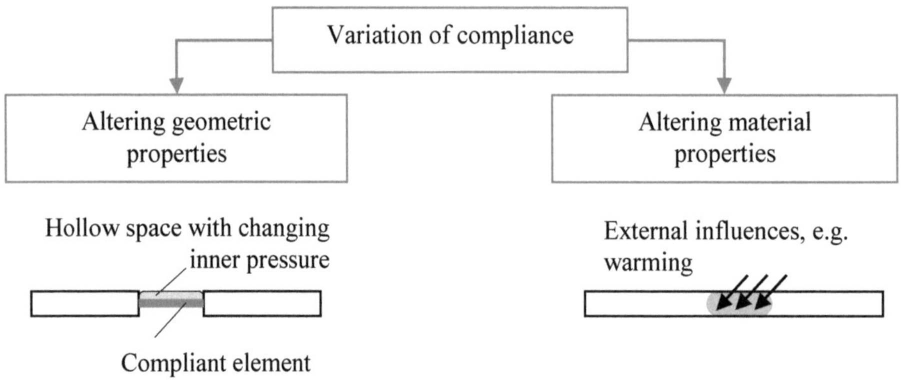

Fig. 2.5: Methods of varying compliance with examples (below)

Changing the compliance of a structure by altering its geometric properties is possible by means of a directed application of energy; for example, by increasing the pressure in a hollow space specifically designed for this purpose. Changing material properties is possible by influencing special materials (functional materials) thermally, electromagnetically or by means of other influences. For example, changing the temperature or warming a specific point of a functional material in a system to alter its mechanical properties. Compliance is subsequently increased locally, thus giving this point the function of a joint.

Using a simple compliant system, the possibility of achieving a specific change in compliance can be discussed. The requirements for this are a geometric linearity, whereby only small displacements are allowable, and a material linearity, where HOOKE's law applies. The system given in the following example is a compliant joint element, which is modeled as a clamped rod with a length l, a YOUNG's modulus E, and an equatorial second moment of area I_z. The rod is loaded at one end with a force $\mathbf{F}$ (Fig. 2.6), while the displacement at the same rod end is given as u_y. The compliance of the system, generally given as $\partial u_y/\partial F$ and here considered as a quotient of the displacement u_y and the force applied, is expressed for $x = l$ as follows:

$$\eta = \frac{u_y}{F}. \tag{2.1}$$

Under the conditions set above, the compliance of the system remains constant. If the elasticity E is linearly dependent on another parameter, e.g. the temperature T, the resultant compliance $\eta(T)$ is also temperature dependent (Fig. 2.6 a). The parameter T can now be used to purposely change the compliance of the system ([120]). Such a system is shown in Fig. 2.7, wherein the increase in temperature of a local point of the system is achieved using a heated filament. The system depicts a tube that is thermally insulated from the filament on one side by a thin silicone layer. The tube, normal "shrink tubing", is first warmed and then sealed at one end in its shrunken form. The rod is then pressurized at the other end. When the filament is heated, it heats a local part on the tube, thus increasing its compliance. Due to the increased pressure inside the tube, the warmed part bends. When the pressure is removed, the system returns to its original state as the warmed material shrinks. The higher the temperature of the local part of the tube, the softer it becomes. The compliance of a system can be purposely varied in this way.

What cannot be guaranteed in such a system is the change in the *sensitivity* of the system $\partial\eta/\partial T$ based on the change in compliance. This remains unchanged for a desired temperature state, because it is already predetermined in connection with the dependence $E(T)$. A dependency $\eta(T)$ is given in Fig. 2.6 b, in which the sensitivity of the system is characterized by the angle α_1. By introducing another parameter, here named X, which should change the elasticity of the material $E(T,X)$, then it is possibly to alter the sensitivity of the system $\partial\eta/\partial T$, based on the parameter T for a particular state, described with T and u_y. The case where the compliance is dependent on two parameters, whereby the change in sensitivity is possible, is schematically presented in Fig. 2.6 c-d. In the interest of simplification, a linear relationship between compliance and the two parameters has been assumed. These expressions apply when the compliance is dependent on the given parameters, which correspond to a function in a mathematical sense.

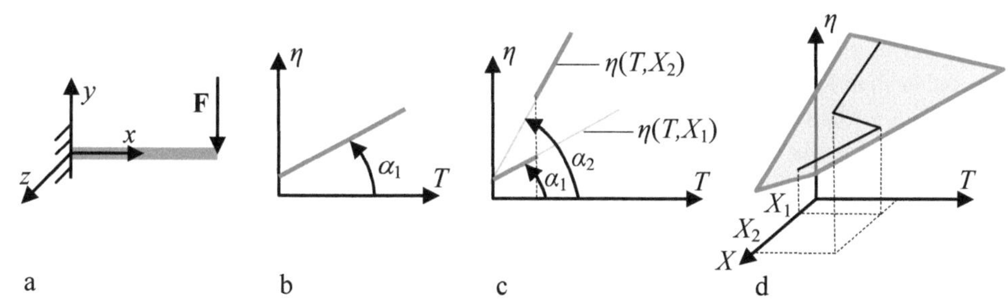

Fig. 2.6: Sensitivity based on the compliance of a system; a – an example of a fixed, loaded rod; b – compliance $\eta(T)$ with invariable sensitivity; c – compliance is dependent on two parameters from $\eta(T,X)$, through which changing sensitivity is possible, characterized by α_1 and α_2; d – spatial depiction of the methods of varying the sensitivity on the area $\eta(T,X)$

The change in sensitivity $\partial\eta/\partial T$ is only possible here if there is a dependence on two or more parameters. This situation can be modeled if the system described in Fig. 2.7 is not used in the role of a fluid-mechanical actuator but as a simple coherent joint, which bends under an external force, as in Fig. 2.6 a. Here there are two available parameters with which to influence the compliance of the system: the temperature, variable using the electrical voltage of the heated filament, and the inner pressure of the tube. In this case, the sensitivity can be altered according to Fig. 2.6 c-d. Here, the inner pressure of the tube takes on the role of parameter X. This example offers an opportunity of varying the sensitivity of a compliant system. Precise alteration of this sensitivity is especially important in measurement technologies.

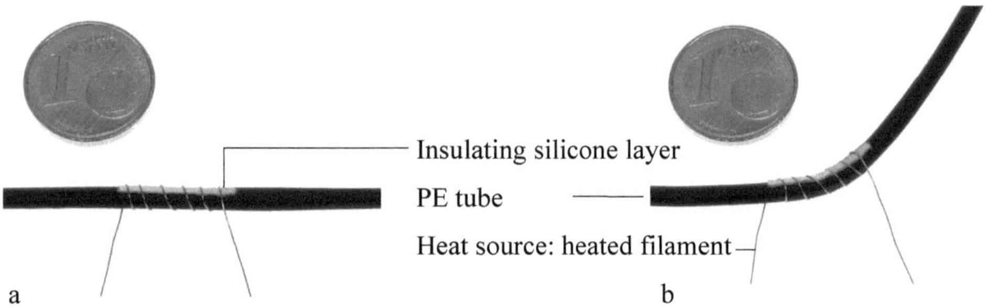

Fig. 2.7: Change in the compliance of a system resulting from temperature variation; a – a tube (shrink tubing) wrapped in a heated filament and with an insulating silicone layer on one side of the tube, no inner pressure applied; b – the system under the effects of temperature increase and inner pressure

2.2 Compliant mechanisms

A mechanism that gains its mobility fully or partially from the compliance of its deformable parts rather than from rigid-body joints only is named a *compliant mechanism* (see also [18], [119]). Due to their compliance, compliant mechanisms offer a number of advantages over rigid-body mechanisms. In general, these can be summarized as follows: less friction or lubrication, good applicability to miniaturization, simple production through less complicated or no assembly and low maintenance. It is, however, important to consider the mostly complicated paths of output elements, that are difficult to describe theoretically, as well as signs of fatigue of material from larger deflections.

2.2.1 Classification of compliant mechanisms

Based on the definition of compliant mechanisms, these can be separated into the categories of fully compliant mechanisms and partially compliant mechanisms (Fig. 2.8). If a mechanism undergoes motion due to the presence of compliant structural sections or joints, as well as from rigid-body joints, this mechanism is named a *partially compliant mechanism*. A *fully compliant mechanism* undergoes motion only due to the compliance of its structural parts.

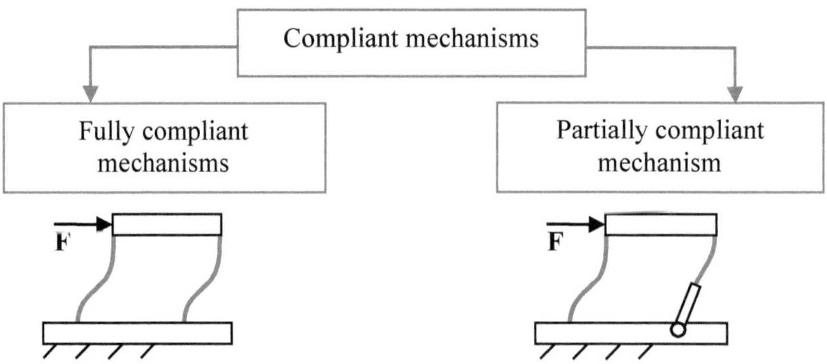

Fig. 2.8: Classification of compliant mechanisms by means of the structural design

Additionally, fully and partially compliant mechanisms can be separated further into mechanisms with concentrated or distributed compliance, depending on the distribution of its compliance (see Section 2.1.1). This results in four fundamental cases of compliant mechanisms, schematically depicted in Table 2.1 using the example of a linear path-generating mechanism with only revolute pairs. In principle, the compliant joints and rigid-body joints can be revolute pairs, prismatic pairs or others.

Table 2.1: Classification of compliant mechanisms by means of the structural design and the distribution of compliance using the example of a path-generating mechanism

Mechanism	Fully compliant mechanism	Partially compliant mechanism
With concentrated compliance	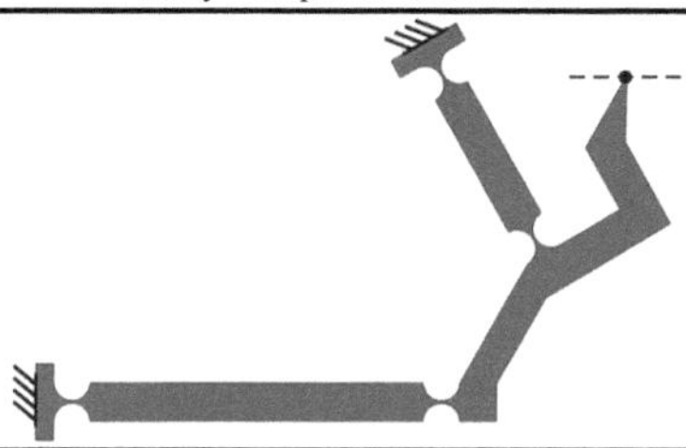	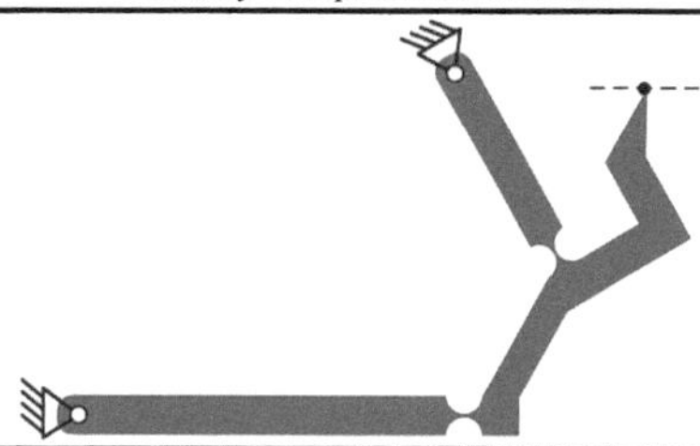
With distributed compliance	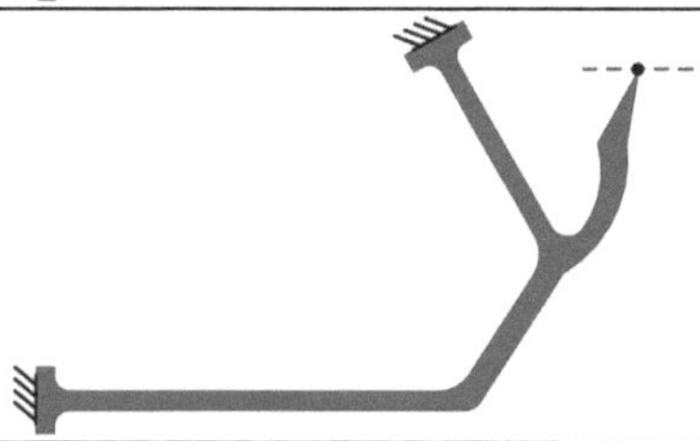	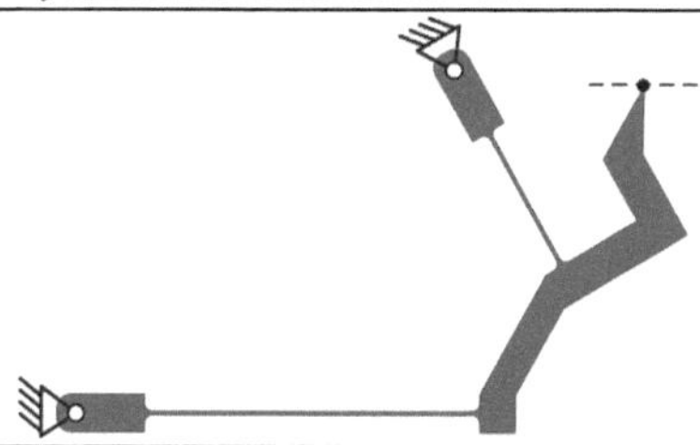

2.2.2 Compliant joints

A structural part of a compliant mechanism with a greatly increased compliance can be seen as a *compliant joint*, which allows at least one relative motion due to deformation, but it is often limited to a localized area. Conversely to rigid-body joints, in which two rigid links form either a form-closed or force-closed pairing, neighboring links of a compliant mechanism are connected to each other in a materially coherent way. Such a pairing is named a *coherent joint*. Motion of the links relative to each other is only possible due to the compliance of their connection. As well as the term coherent joint, other terms for a coherent pairing do exist that are more specific, such as a *flexure hinge* (see Chapter 6).

Such a joint in a compliant mechanism can be achieved in two ways. Firstly, an increased compliance can be achieved through variation of the geometry, that is a geometric variation in a local area (Fig. 2.9 a). Secondly, a joint can be designed that exhibits increased compliance due to the implementation of another material in the system (Fig. 2.9 b). A combined variation is also possible, in which a joint is given increased compliance because of a geometric alteration and use of other materials. The geometric design of the joints is based on certain parameters (Fig. 2.9 a), such as the length of the joint l, as well as its geometric profile $y = y(x)$ (see [58]-[60]). A classification of the structural sections into compliant and rigid parts, and thus an abstraction of these as joints and links, often simplifies the design of a model of such a compliant system (see also Chapter 6). If such a separation is not possible, the compliant system should be modeled as a fully compliant system, as in [40] and [41].

Table 2.2: Classification of compliant joints by means of different joint characteristics

Criterion	Category	Subcategory	Example
Cause of relative motion	Compliance	Variation of geometry	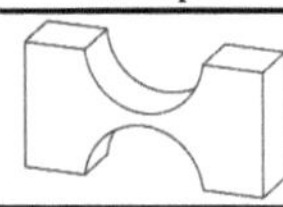
		Variation in material	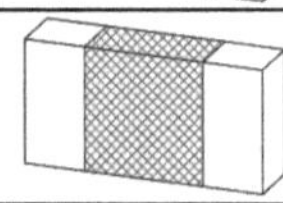
		Variation of geometry and material	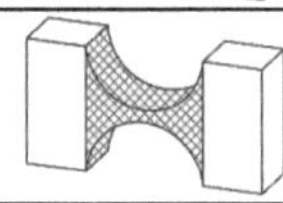
	Compliance and form or force closure		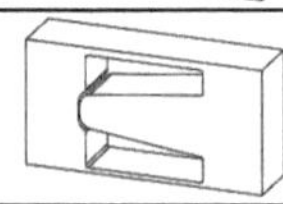
Distribution of compliance	Concentrated		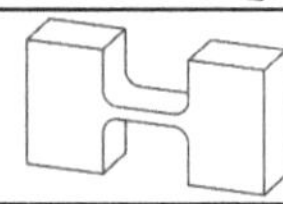
	Distributed		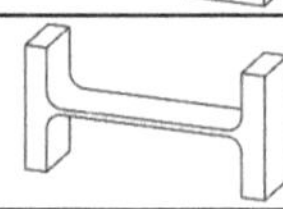
Structural design/ assembly	Monolithic/one-part		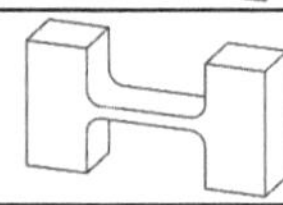
	Non-monolithic/multi-part		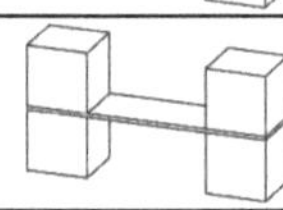
Geometry of solid body	Prismatic		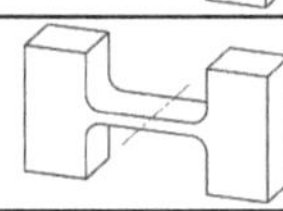
	Non-prismatic	Rotation-symmetric	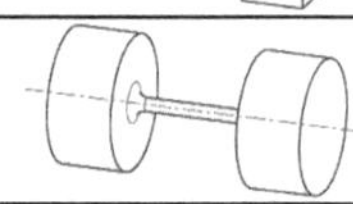
		Plane-symmetric	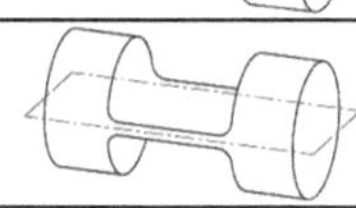
		Asymmetric	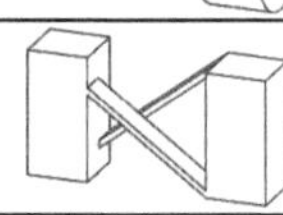
Complexity of joint	Low		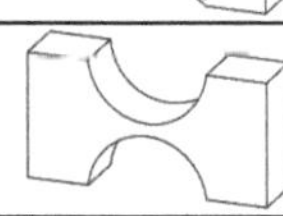
	High		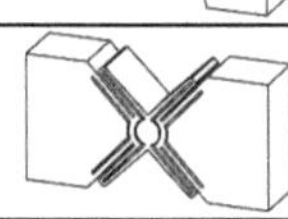

Table 2.3: Classification of compliant joints by means of different joint properties

Criterion	Category	Subcategory	Example
Desired degree of freedom of joint (f)	$f = 1$		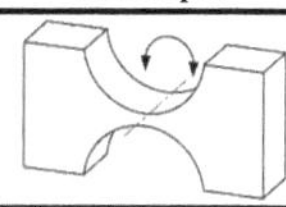
	$f = 2$		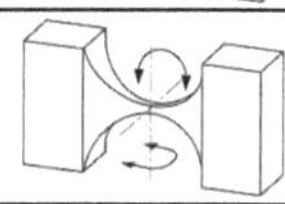
	$f \geq 3$		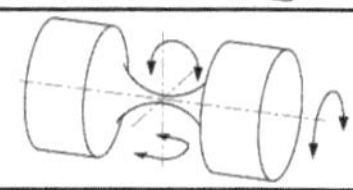
Form of relative motion (for $f = 1$)	Rotation		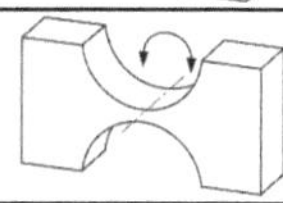
	Translation		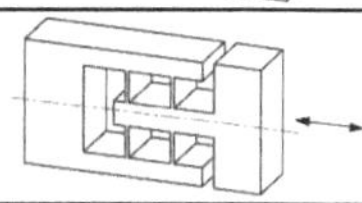
	Screw motion		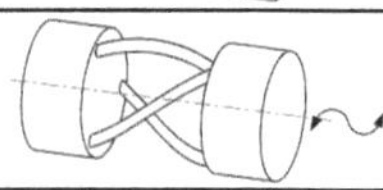
Position of joint axes (for $f = 2$)	Planar	Parallel axes	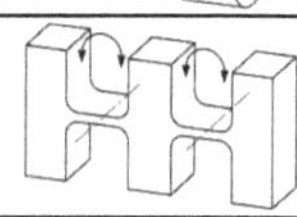
		Coincident axes	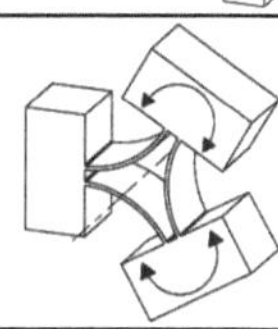
	Spherical		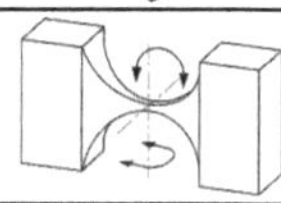
	Spatial		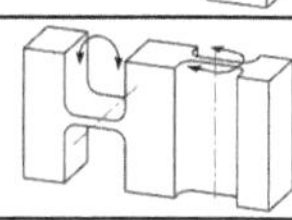
Kind of stress (for rotational motion)	Bending		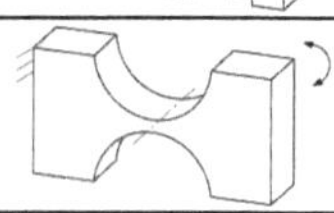
	Torsion	Rod axis parallel to x-axis	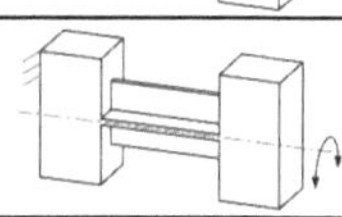
		Rod axis parallel to z-axis	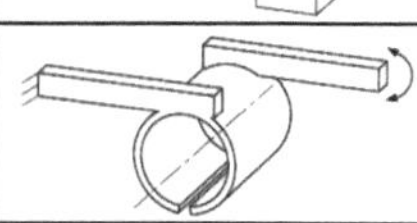
	Bending and torsion		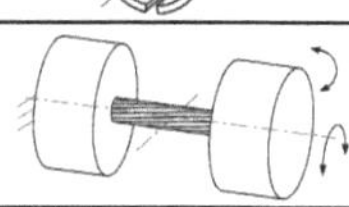

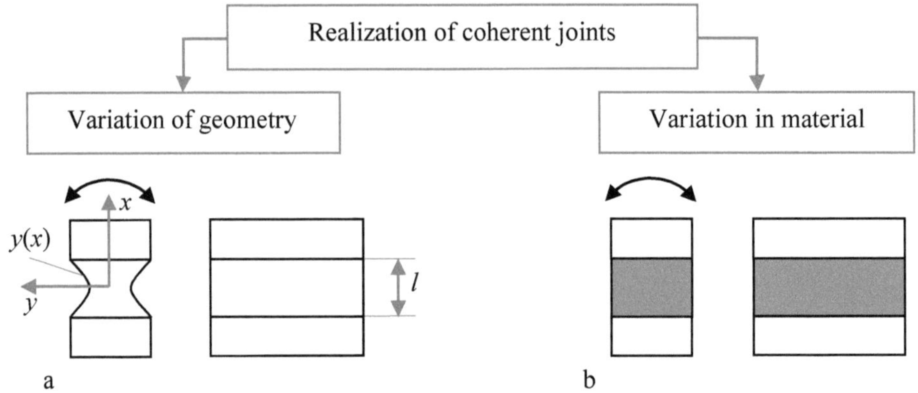

Fig. 2.9: Coherent joint of a compliant mechanism from two perspectives, realized through; a – design by variation of geometry; b – design using a material with increased compliance

Compliant joints in the used mechanisms mostly realize the function of revolute pairings with the *degree of freedom f* from one, although compliant prismatic pairings have also been suggested in relevant literature (e.g. ([19], [97]). There are a number of further criteria for classifying compliant joints (Table 2.2 and Table 2.3; see also [58]). Similar to the classification of rigid-body mechanisms, compliant joints can also be classified using various criteria. The following classification makes a distinction between joint characteristics (Table 2.2) and joint properties (Table 2.3). Due to the inherent shift of the rotational axes of compliant revolute joints (see also Section 6.2), the specification of the two properties "degree of freedom" and "form of the relative motion" is only possible in an idealized manner. This classification reveals the numerous possibilities of designing compliant joints for use in planar, and especially in spatial compliant mechanisms.

As before, there are few guidelines, theoretical resources or design tools for the systematic development even of simple compliant joints. The first steps of synthesizing compliant mechanisms, based on the *rigid-body model*, as well as the purposeful design of coherent revolute joints by using suitable or optimized contours in notch flexure hinges will be described in Chapter 6 of this book.

Compliant systems with joints formed through variation of geometry or variation in material can also be found in nature. One example of joints that are only formed through geometric design is multi-jointed spider leg. A spider's leg is coated in a relatively hard shell, a so-called exoskeleton. It also has flexible formations with foldable membranes, wherein all areas are coherent and consist of the same material (cuticle). When the pressure inside the shell increases, the membrane unfolds and the leg is stretched (Fig. 2.10).

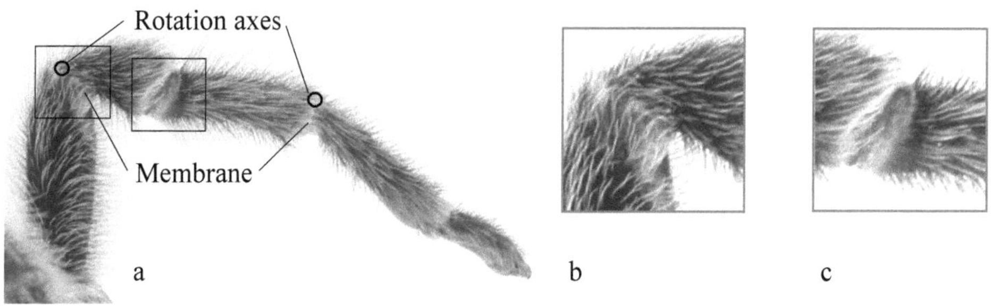

Fig. 2.10: An example of a compliant system in nature; a – the extension of a spider's leg is realized increasing the hydraulic pressure in the joints; b – a purely hydraulic joint, the axis of rotation passes through a peripherical point of the leg; c – a joint which is extended only using muscles, the axis of rotation intersects the cross-section area of the leg

Three cases for the spider's leg extension are classified in dependence of geometry of the joint area ([118]). The first case is a joint which realizes the extension of the leg solely due to the increase of hydraulic pressure. In this case, the axis of rotation passes through a peripherical point of the leg and the participation of muscles in the extending motion is thus excluded (Fig. 2.11 a). Another joint enables the leg extension due to the increase of pressure in the joint area with the participation of muscles at once, because the axis of rotation intersects the cross-section area of the leg, as depicted in Fig. 2.11 b. In the third case, the membrane surface is divided into two equal areas by the axis of rotation, so that the hydraulic pressure causes equal opposite acting moments with respect to the axis of rotation. In this case, no motion results (Fig. 2.11 c). Thus, the leg is extended or bent only using muscles.

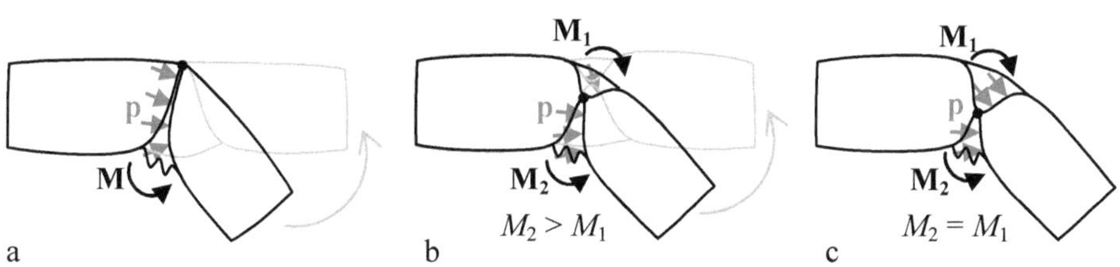

Fig. 2.11: Schematic depiction of spider's leg; a – a purely hydraulic joint; b – a combined joint, joint is extended hydraulically and using muscles; c – a joint, in which the joint is extended only using muscles

The spinal column, consisting of vertebrae and intervertebral discs, is another example of a biological joint that is only possible due to additional materials with increased compliance. Between the vertebrae of the spinal column are the intervertebral discs, which consist of an outer fibrous ring and an internal gel-like material (Fig. 2.12). The intervertebral discs are connected to the base of the spinal column with a fibrous connective tissue.

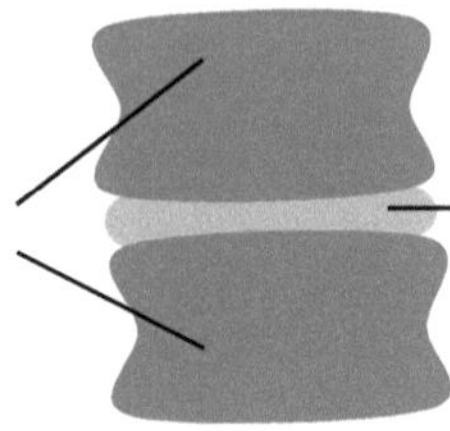

Fig. 2.12: Schematic diagram of two vertebrae of the spinal column; vertebrae are connected via a intervertebral disc, the elastic disc forms a joint

The intervertebral discs with their higher compliance allow the vertebrae to move relative to one another and thus grants the entire spinal column a certain flexibility. Both examples deal with coherent joints, because there is a coherent connection between the compliant joint elements and the other parts of the system.

2.3 Compliant actuators and sensors

An important condition for actuators and sensors, which are to be used in compliant systems, is that these should also be compliant and therefore not affect the mechanical functionality of the system as a whole. Generally, the sensory and actuatory properties of a compliant system are realized either geometrically or materially and it is advantageous when these system components also represent indispensable, mechanical components of the system, that is, components that are coherent connected to the system as a whole. In such cases, this results in a multi-functionality of those system components.

2.3.1 Compliant actuators

If a compliant system or a body is not considered to be a kinematic structure, but the cause of movement in other structures, then the term actuator can be used (see also [9]). A *compliant actuator* is an actuator that performs a mechanical action through the deformation of one or more compliant parts (Fig. 2.13). If it is the energy of a fluid medium (gas, liquid or gel) that is transferred into mechanical work (deformation), such an actuator is named a *fluid-mechanical compliant actuator* (Fig. 2.13 a).

The example of spider leg joints, shown above in Fig. 2.11 a-b, also dealt with a fluid-mechanical actuator used in the joint. The system in Fig. 2.11 c is not a fluid-mechanical actuator, because the movement is only elicited by use of muscular strength.

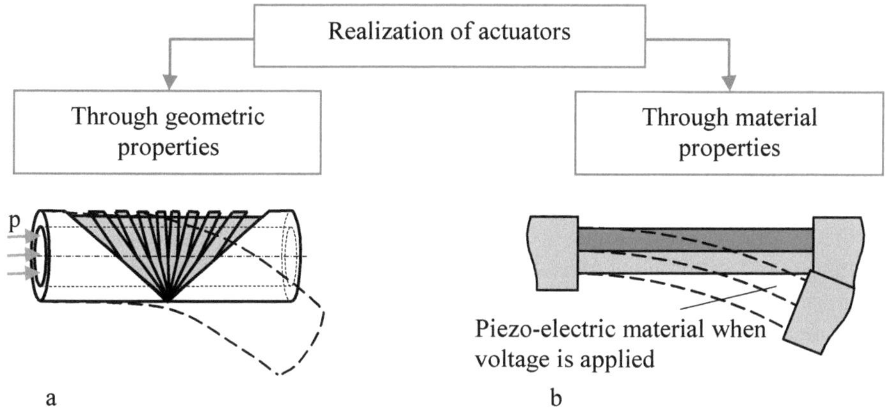

Fig. 2.13: Schematic depiction of compliant actuators; a – a fluid-mechanical actuator, which bends due to an increase in internal pressure; b – due to the use of functional materials, e.g. piezo-electric materials, that change their length when an electrical voltage is applied

A further example is a compliant gripping device, with fluid-mechanical actuators arranged in pairs parallel to one another, as shown in Fig. 2.14 a-c. The gripping device can be used to grip either inwards or outwards due to the arrangement of the individual gripping fingers ([121]). Such a gripping mechanism can be seen as a compliant mechanism with an inherent actuating function. The actuator properties are possible because of the hollow spaces (structural and geometric) and the elasticity of the material, that makes up an indispensable part of the mechanism. Other examples of fluid-mechanical compliant actuators can be found in [36]-[38], including one example in combination with a rigid-body mechanism in [28].

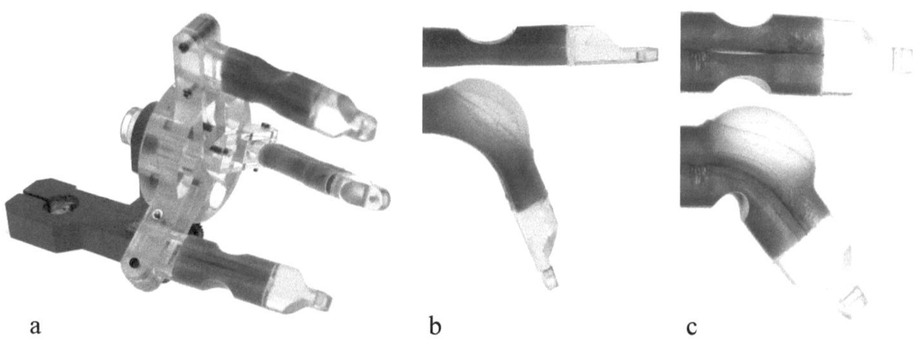

Fig. 2.14: A gripping device; a – a gripper as a compliant mechanism with fluid-mechanical actuators ([121]); b – a fluid-mechanical actuator; c – fluid-mechanical actuators arranged parallel to one another

2.3.2 Compliant sensors

One example for the realization of a *compliant sensor*, an elastically deformable sensor, would be a compliant gripping device, made from a polymer with inherent sensory properties (Fig. 2.15 a). Here, a difference will be drawn between the states "with object", if an object is held by the gripper, and "without object", if no object is held between the compliant gripping elements. For this example, polymers will be used that conduct electricity. These materials possess almost the exact same mechanical properties as the conventionally-used polymers used to produce the gripping device. Conductive polymers alter their electrical resistance depending on the mechanical load applied to them. The electrical resistance decreases under a compressive load and increases under a tensile load. On a first level of abstraction, these properties can be accounted for by the change in distance between the conductive particles in the polymer, graphite particles, for example. Conclusions regarding the mechanical stress in the material can be made using the change in electrical resistance, two states can be easily distinguished from one another in this way ([50], [51]). Firstly, the mechanical stress σ_1 in the material for the "without object" state and the stress σ for the "with object state" should be compared. Subsequently, the positions of the compliant elements with the greatest discrepancy $\sigma - \sigma_1 = \sigma_2$ are identified, in order to select suitable conductive polymers for the realization of the sensory applications (Fig. 2.15 b).

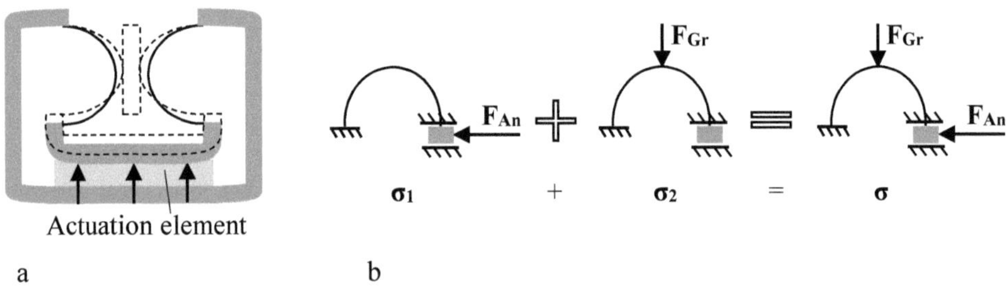

Fig. 2.15: Example of a compliant gripping device; a – gripping positions in relaxed state, and under application of a tractive force with a gripped object; b – outline of the superposition principle for the stress in the compliant element

In order to derive the normal stresses, linear theory is applied; with the assumption that the deflections remain small. As a result, the superposition principle can be used, whereby the method of finding the stress differential σ_2 is greatly simplified. The mechanical stress σ_2 in the compliant element can be found under the effect of only one gripping force. It can be concluded from the symmetry of the system that the reaction forces and moments at the right and left ends have the same value. These values are named F_R for a force and M_R for a moment (Fig. 2.16 a), whereby the following applies for the reaction forces:

$$F_R = \frac{F_{Gr}}{2}. \qquad (2.2)$$

The internal forces, moments and ultimately also the mechanical stresses are determined according to the parameter φ (Fig. 2.16 b). The resulting internal forces and an internal moment are the normal force N_S, that acts perpendicular to the cross-sectional plane, the lateral force Q_S and the internal bending moment M_S. The equilibrium conditions for a segment of the compliant element (Fig. 2.16 b), whereby the moment equations are formulated according to their coordinate origin, provide the values for the internal forces and an internal moment for the interval $0 < \varphi < \pi/2$. Due to the symmetry, the values of the forces and the moment for the interval $\pi/2 < \varphi < \pi$ will be the same, although here the y-axis serves as the axis of symmetry.

$$N_S = -\frac{F_{Gr}}{2}\cos\varphi,$$
$$Q_S = -\frac{F_{Gr}}{2}\sin\varphi, \tag{2.3}$$
$$M_S = M_R - \frac{F_{Gr}R}{2}(1-\cos\varphi)$$

The reaction moment can be calculated using *MENABREA's method* (see also [30], [80]). Because only the internal moment M_S is dependent on the reaction moment M_R, the expression for the derivative of the deformation energy W is reduced to only a single integral:

$$\frac{\partial W}{\partial M_R} = \frac{1}{EI_{\varsigma 1}}\int_0^{\frac{\pi}{2}}\left(M_R - \frac{F_{Gr}R}{2}(1-\cos\varphi)\right)d\varphi = 0\,. \tag{2.4}$$

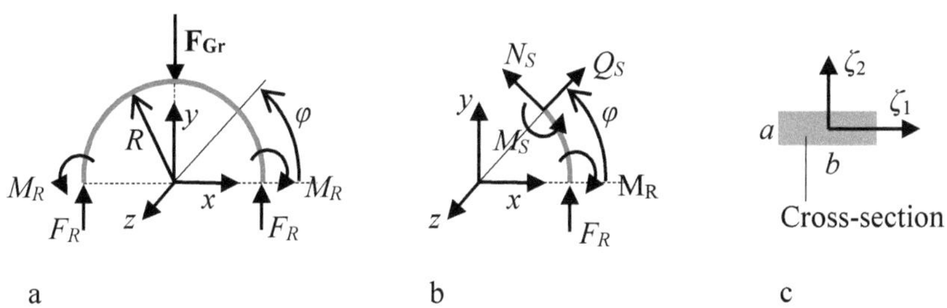

Fig. 2.16: Compliant element; a – support reactions in the compliant element; b – outline of a cutting part of the compliant element; c – cross-section of the compliant element

$EI_{\zeta 1}$ is the bending stiffness of the compliant element (Fig. 2.16 c), which consists of the YOUNG's modulus E and the equatorial second moment of area $I_{\zeta 1}$. R is the radius of curvature for the compliant elements. From equation (2.4), the reaction moment can be found:

$$M_R = F_{Gr}R\left(\frac{1}{2}-\frac{1}{\pi}\right). \tag{2.5}$$

The mechanical stress σ_2 in the cross-section of the compliant element consists of the compressive/tensile stress and the bending stress, which lie orthogonal to cross-sections with area A and the dimensions a and b. A coordinate system with axes ζ_1 and ζ_2 (Fig. 2.16 c) is oriented with the principal axis of the cross-section, so that the axes ζ_1 and z point in the same direction.

$$\sigma_2(\varphi,\zeta_2) = \frac{N_S}{A} + \frac{M_S}{I_{\varsigma 1}}\zeta_2 = -\frac{F_{Gr}}{2ba}\cos\varphi + \frac{6F_{Gr}R}{ba^3}\zeta_2\left(\cos\varphi - \frac{2}{\pi}\right) \tag{2.6}$$

The extreme of σ_2, which coincides with its boundaries $\varphi = 0$ and $\varphi = \pi/2$, can be found using the derivative of the previous equation with respect to φ. With respect to the parameter

ζ_2, this provides a linear dependence. For these reasons, only the values of the tension σ_2 within the boundaries for φ and ζ_2 are compared:

$$\begin{aligned}
&\sigma_2\left(0,-\frac{a}{2}\right)=-\frac{F_{Gr}}{2ba}-\frac{3F_{Gr}R}{ba^2}\left(1-\frac{2}{\pi}\right),\\
&\sigma_2\left(0,\frac{a}{2}\right)=-\frac{F_{Gr}}{2ba}+\frac{3F_{Gr}R}{ba^2}\left(1-\frac{2}{\pi}\right),\\
&\sigma_2\left(\frac{\pi}{2},-\frac{a}{2}\right)=\frac{6F_{Gr}R}{\pi ba^2},\\
&\sigma_2\left(\frac{\pi}{2},\frac{a}{2}\right)=-\frac{6F_{Gr}R}{\pi ba^2}.
\end{aligned} \tag{2.7}$$

The two previous equations give the maxima of the absolute values of the stresses, provided that the radius R is far larger than dimension a. These maxima originate on the cross-section with the coordinate $\varphi = \pi/2$. This involves the underside for $\zeta_2 = -a/2$, which undergoes a tensile stress, and the upper side for $\zeta_2 = a/2$, which experiences a compressive stress. Taking into account the fact that the body held in the gripper exerts a direct pressure on the upper side of the gripping element at $\zeta_2 = a/2$, then even higher stresses must also be included in the calculations. Thus, the point with coordinates $\varphi = \pi/2$, $\zeta_2 = a/2$ (last expression from (2.7)) is chosen for the sensing. The part about this point is made from an electrically conductive polymer and subjected to an electrical voltage. By measuring the change in resistance, it can be determined if an object is being gripped or not (Fig. 2.17 a-b).

When the middle layer of the compliant element is to be equipped with a sensor, the point $\varphi = \pi/2$ cannot be used to insulate the electrical voltage of the object from the surroundings, because the voltage σ_2 at point $\zeta_2 = 0$ has the value of zero. Applying $\zeta_2 = 0$ to the mechanical stress from equation (2.6) results in an expression for the mechanical stress of the middle thread or strip of the compliant element:

$$\sigma_2(\varphi,0)=-\frac{F_{Gr}}{2ba}\cos\varphi\,. \tag{2.8}$$

The point $\varphi = 0$ and thus also the point $\varphi = \pi$ is important for the design of the sensory properties of a system (Fig. 2.17 a).

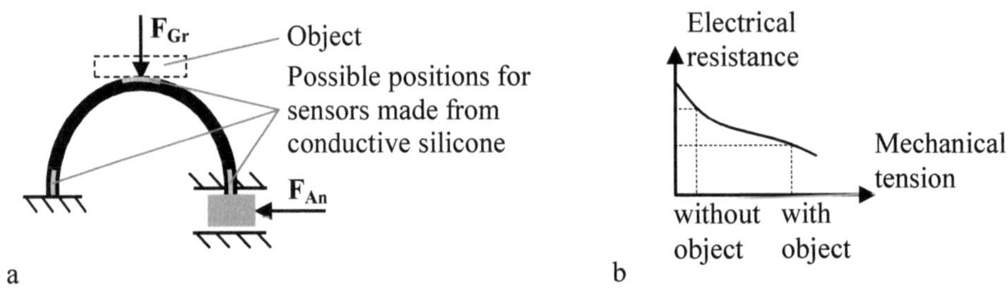

Fig. 2.17: Use of electrically conductive polymers; a – compliant element of a gripping system with possible positions for the integration of inherent sensing capabilities; b – depiction of a qualitative dependence of electrical resistance of the mechanical stress for electrically conductive polymers that contain graphite particles

Electrically conductive and non-conductive polymers can be seamlessly combined together, so that they form a coherent body. Electrically conductive polymers possess almost exactly the same mechanical properties of non-conductive polymers. Therefore, the position made from electrically conductive polymers take on the sensing functions of the system in the application described above while remaining a component part of the mechanical, compliant system. Additionally, the part where the gripping elements have contact with the gripped object can be made from an electrically conductive polymer with the same or lower SHORE hardness value than the rest of the system parts, the point $\varphi = \pi/2$ is chosen for the sensing functions of the system. In this case, this system part also takes on the function of softly gripping the object, while also increasing the level of multi-functionality yet further. Further examples of using electrically conductive polymers can be found in [14] and [15].

2.3.3 Multi-functionality

As a result of their compliance, compliant mechanisms can be designed with inherent actuator and/or sensory properties. An *inherent property* of a system is one that was attained through mechanical means, be they structural, geometric or stemming from the properties of the material used. Conversely, conventional rigid-body mechanisms are moved using actuators that are often only spatially integrated into the system as a whole. If such a rigid-body system should contain a sensor, this is generally added as an additional part to the system. Compliant systems offer the opportunity to use inherent sensors, by choosing materials with special compliant properties. Thus a subsystem takes on multiple functions and increases the level of multi-functionality of the system. *Multi-functionality* is understood as the ability of fewer components, or even one single component, to take on multiple functions.

Fig. 2.18 gives a comparison of a traditional rigid-body motion system and a multi-functional compliant system on an abstract level. A rigid-body motion system (Fig. 2.18 a) consists of a mechanical subsystem of actuators, which allow for motion of the rigid bodies relative to one another, and sensors, which monitor the condition of the system. A compliant system can, due to the use of materials with functional properties, take on the functions of the actuators and sensors. Such systems exhibit multi-functionality due to a deliberate structural and geometric design that eliminates the need for further sensor components. This results in a compliant system with inherent actuator and sensory properties. It is even possible to reach maximum multi-functionality by having the sensor and actuator systems also form the main mechanical components of the system (Fig. 2.18 b).

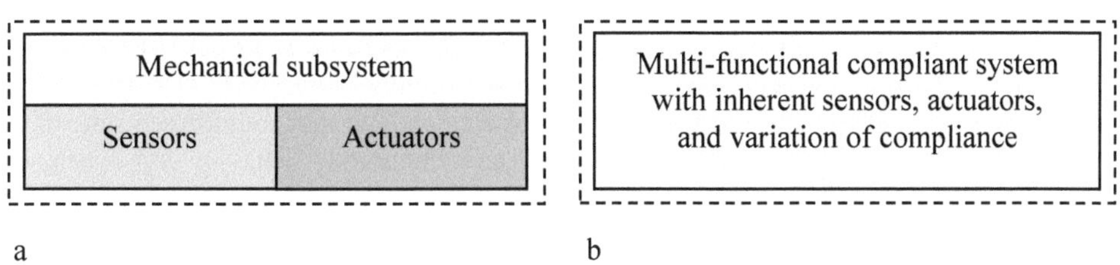

Fig. 2.18: A comparison of two systems; a – schematic depiction of the fundamental components of a rigid-body motion system; b – depiction of a compliant motion system; the mechanical subsystem takes on the function of individual components

2.4 Motion behavior of compliant systems

An additional criterion for the classification of compliant systems is the *motion behavior* ([120]). According to the NEWTON-EULER equations, the sum of all acting forces or moments induces a change in the motion, which is mirrored by the derivative of the impulse $m\mathbf{v}$ or the angular momentum $\mathbf{D}$ with respect to time t. The following applies for a rigid body:

$$\sum_{i=1}^{I} \mathbf{F}_\mathrm{i} = \frac{d(m\mathbf{v})}{dt},$$
$$\sum_{j=1}^{J} \mathbf{M}_\mathrm{j} = \frac{d\mathbf{D}}{dt}. \tag{2.9}$$

When the velocity remains very small or only changes by a small degree, dynamic effects represented on the right side of the equation can be disregarded. This is assumed for the classification of the motion behavior of compliant systems. Each force or moment that increases from zero is contrasted with the corresponding position of an end effector element of the compliant system.

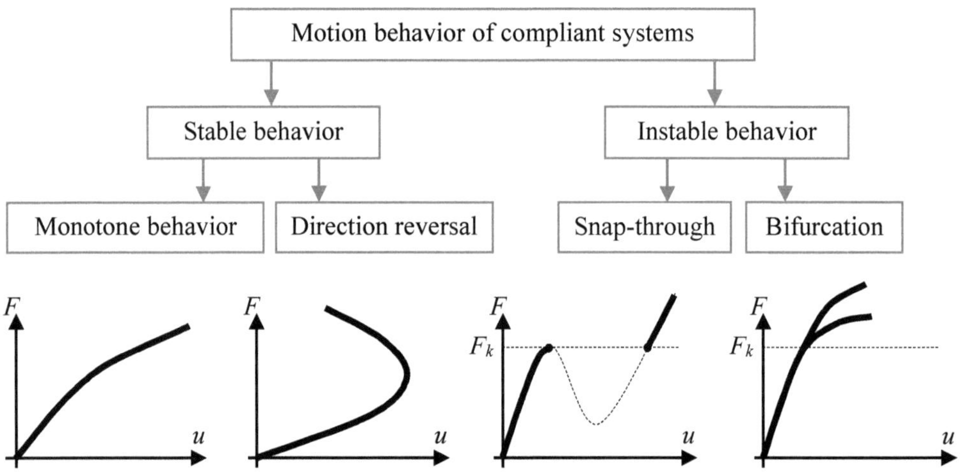

Fig. 2.19: Classification of the motion behavior of compliant systems (see also [120])

Motion behavior is subdivided into stable and unstable behaviors (Fig. 2.19). In cases of stable motion behavior, a certain load, given the name F, only corresponds to a parameter u, which characterizes the position of the end effector element and thus a deformed state of the system. Furthermore, a distinction is made between a monotone behavior and a behavior with a reversal of direction. With an unstable behavior of the compliant system, a snap-through effect (motion behavior with a "step") and a bifurcation (a "branch" of the behavior) may be observed. The types of motion behavior will be considered using specific examples in following sections, although rigid-body mechanisms as well as compliant systems will be used to show the differences between the transferal of motion using both rigid-body and compliant systems. It should be noted here that the classification of a system into one of these categories is dependent on the choice of the defining parameter u.

2.4.1 Stable motion behavior

Most compliant systems exhibit *stable motion behavior*. This behavior is essential in precision technology, robotics, gripping technology and in numerous other areas, such as moving a positioning stage or designing a specific path for the precise manipulation of workpieces.

Monotone behavior

In Fig. 2.20, different examples of *monotone motion behavior* are given: a rigid-body mechanism, here an inverted slider-crank mechanism; a compliant mechanism, which moves based on the principle of a parallel-crank, and a fluid-mechanical actuator.

As the input size increases, the characteristic sizes that constitute the output motion also increase. In the example of the rigid-body mechanism, an inverted slider-crank mechanism, its input motion is defined by the angle of the element that completes a full rotation. As the input angle of the crank increases, the output angle will also increase in a monotone way.

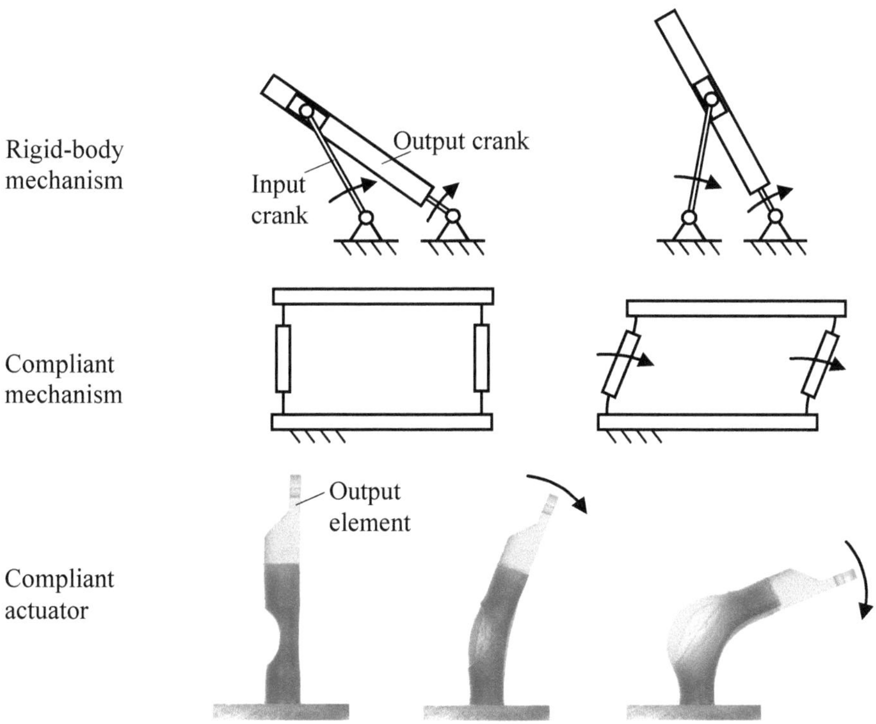

Fig. 2.20: Monotone behavior; a rigid-body mechanism as an inverted slider-crank mechanism, a compliant mechanism and a compliant actuator; as the input angle increases, so does the output angle; compliant actuator with an output element moves when pressure is increased in the hollow space of the fluid-mechanical actuator

The motion ranges for the input and output links of a compliant mechanism are limited. Within these ranges, the input angle increases as the output angle increases, although one of the frame-mounted links can be seen as an input link and the other as an output link.

In the case of an actuator, a monotone motion is characterized as motion where increasing inner pressure corresponds to a growing inclination angle of the output element mounted on the actuator. The motion range of the output element is limited in this case.

Behavior with a direction reversal

A *motion behavior with a direction reversal* can be used to design complex motion paths. Fig. 2.21 shows a rigid-body mechanism, a compliant mechanism and a compliant fluid-mechanical actuator. The slider of a slider-crank mechanism that forms a rigid-body system, carries out an alternating motion – motion with a direction reversal – while the input direction remains the same. The *motion range* of the compliant system is limited; as the input link moves in one direction, the output link undergoes a direction reversal. The geometry of the actuator is optimized in such a way, that the output element undergoes a direction reversal when a load is applied as the inner pressure increases. Displacement can be chosen as a characteristic motion value for the slider-crank mechanism, and in the other two cases, an angle of inclination of an output element can be used.

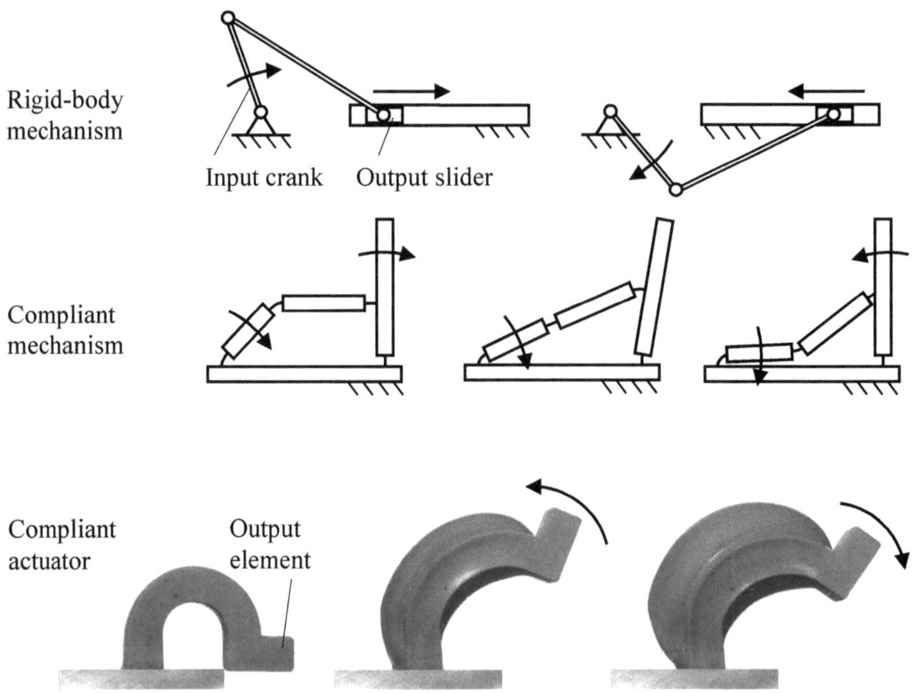

Fig. 2.21: Behavior with a direction reversal; a rigid-body mechanism as a slider-crank mechanism, a compliant mechanism and a compliant actuator; an input motion is applied on the left, the output link on the right changes its direction, while the input link maintains its direction; compliant actuator with an output element changes its motion direction when the pressure in the hollow space of the fluid-mechanical actuator is increased

The property of a direction reversal can be used, for example, in the case of an actuator (Fig. 2.21), for locomotion tasks, due to the specific path the components takes. In such applications, the corresponding structures can be used as extremities or as component parts of a motion system.

Compliant systems that exhibit behavior with a direction reversal have another noteworthy property. A characteristic output value can correspond to two input loads (pressure, force or moment) in a particular area (Fig. 2.22 a). In this way, a system can reach two different compliances for a position u^*, described with the same parameter.

A motion behavior that lies between monotone behavior and behavior with a direction reversal corresponds to a limited growth of the characteristic motion value of the output element, while the driving parameter increases steadily (Fig. 2.22 b). Rigid-body mechanisms such as stepping gears, i.e. a Geneva drive, exhibit this property. During the dwell phase of the output element, the driving link continues to move. Compliant mechanisms can also exhibit such behavior, for example by using an end stop for the output element. Another method of achieving behavior of this kind is purposefully designing the geometry of a compliant system to this end. Fig. 2.22 c depicts a compliant actuator made of silicone, whose characteristic displacement u undergoes no further change after a certain load is reached. This behavior was achieved by using model-based optimization. Such an actuator is very advantageous, for example, when designing a gripping system. This property eliminates the sensory effort required to monitor the gripper force, it is limited only by the mechanical properties of the actuator.

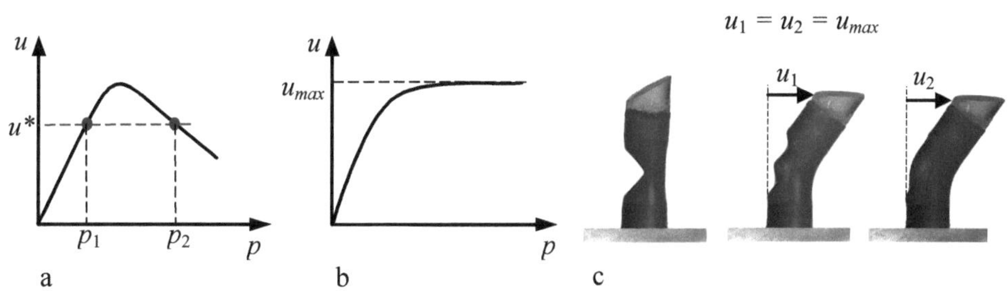

Fig. 2.22: A motion behavior between monotone behavior and behavior with a direction reversal; a – behavior with a direction reversal: a position u^* can be reached using loads of either p_1 and p_2, both positions exhibit different values of compliance; b – behavior between monotone behavior and behavior with a direction reversal; c – a fluid-mechanical actuator with a behavior as depicted in b

2.4.2 Unstable motion behavior

As opposed to stable motion behavior, a compliant system exhibiting *unstable motion behavior* can assume various different positions under a given load. Unstable motion behavior can appear as snap-through or as bifurcation.

Behavior with a snap-through effect

An abrupt transition from one equilibrium position into another can be named a snap-through, thus a critical load can result in a number of different equilibrium positions. *Motion behavior with a snap-through effect* can often be seen as the result of the compliance of a system. A simple example of this effect is an elastic beam clamped at both ends, where the distance between the two clamped points is smaller than the total length of the beam (Fig. 2.23 a). For this reason, the beam shows an equilibrium position with a curvature. When a growing force acts on the middle of the beam, the beam sudden snaps to the other equilibrium position due to a critical load. Thus, the beam has two equilibrium positions in an unloaded state. In this case, this behavior is symmetrical.

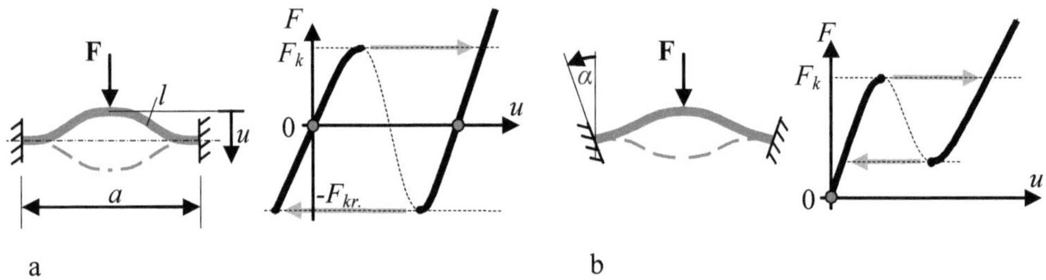

Fig. 2.23: A beam clamped at both ends as an example for a compliant system exhibiting behavior with a snap-through; a – a system with two symmetrical equilibrium positions for an unloaded state $F = 0$; b – a system with a single equilibrium position for an unloaded state when $F = 0$

If the clamping points are inclined towards one another by an angle 2α, the symmetrical properties of a horizontal symmetry axis are eliminated, which affect both the geometric design of the system, as well as its behavior. When the inclination angle reaches a certain point, the system only has one equilibrium position for an unloaded state, as depicted in Fig. 2.23 b.

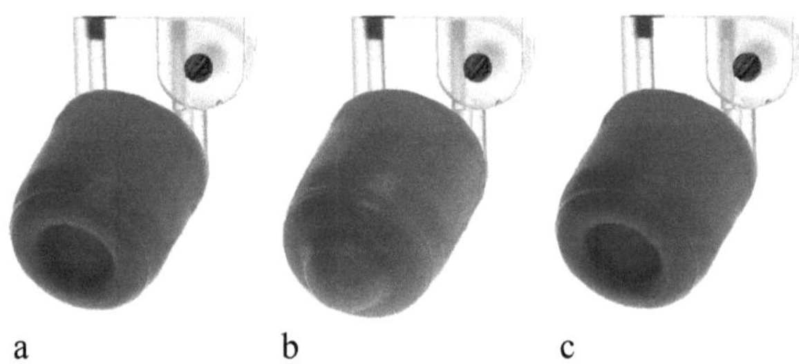

Fig. 2.24: Snap-through of a curved structure with monostable motion behavior; a – original position; b – position after application of a critical pressure; c – position after removal of pressure, this corresponds with the original position

Fig. 2.24 and Fig. 2.25 show rotationally symmetrical structures that, in their original positions, have a semitoric bulge around a spherical, inverted dome. When the inner pressure is increased, the spherical dome is pushed outwards; this motion also has a snap-through effect, which can be utilized as a fluid-mechanical actuator. The characteristic feature of such actuators is either a monostable or bistable motion behavior that can be precisely altered by choice geometric parameters such as radii, wall thickness among others. The outer dimensions of the actuators in Fig. 2.24 and Fig. 2.25 are the same. The thickness of the walls of the first actuator (Fig. 2.24), however, is higher, while the second actuator has thinner walls. The critical load has different values for each of the two systems. When this critical load is reached, an arbitrarily small load results in a very large displacement and the dome is pushed completely outwards. If the pressure is removed, the first structure (Fig. 2.24) returns to its original position, thus exhibiting a monostable motion behavior, which is also depicted in Fig. 2.23 b.

The second actuator (Fig. 2.25) assumes another stable equilibrium position after the pressure is removed, thus this actuator exhibits a *bistable motion behavior*, which also applies to the system described in Fig. 2.23 a. One possible application of such structures is as mechanical valves ([90]), where the dome is moved under a critical pressure and provides either an opening or a closure (see also [89]).

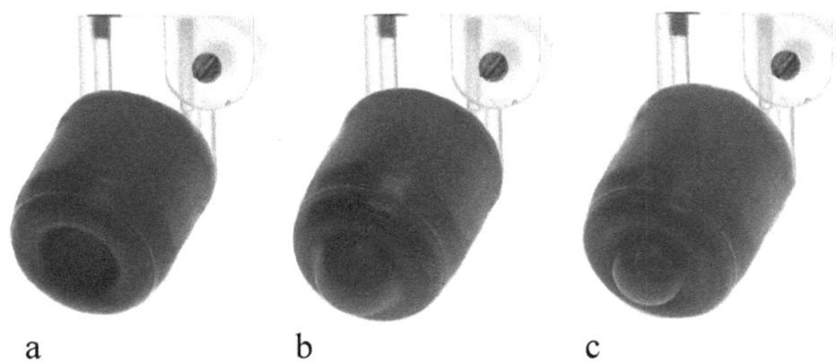

Fig. 2.25: Snap-through of a curved structure with bistable motion behavior; a – original position; b – position after application of a critical pressure; c – position after removal of pressure, the system rests in a new equilibrium position

A further example deals with a compliant actuator with two rotationally symmetrical bulges and a dome. Increasing the inner pressure first leads to a deployment of the bulge and then the middle dome. When the pressure drops, first the dome and then the bulge retracts (Fig. 2.26).

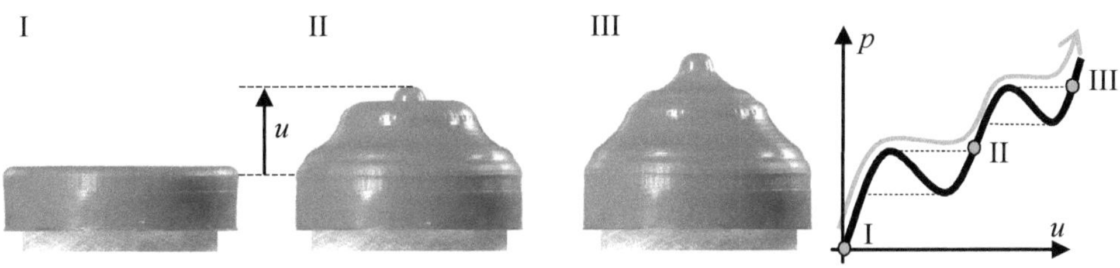

Fig. 2.26: A compliant fluid-mechanical actuator with multiple snap-through effect and the corresponding compliance between the inner pressure and the displacement u of the middle dome point upwards (see also [38])

If the walls are produced thin enough, the actuator can exhibit a *multi-stable snap-through behavior*. Thick-walled structures with less pronounced curvatures result in monotone behavior without a snap-through effect. Certain wall thicknesses can result in mixed behavior, where the dome exhibits a snap-through but the bulge does not. Attaining a compliant structure with specific behavior is possible by choice of certain geometric and material parameters, as well as by changing the material properties of the system due to use functional materials.

Bifurcation

Motion behavior with bifurcation is a behavior with branches: after a critical load is reached, multiple random patterns of behavior are possible. The well-known example of this behavior is the buckling of EULER of a straight rod. Situations where the application of a load to a structure leads to a bifurcation, are often deliberately avoided in technological applications. The following discussion provides possibilities of using this behavior to the advantage of technical systems.

As examples of bifurcation behavior, structures under an axial load are considered. Geometric structures of this kind are shown in Fig. 2.27 (see also [115]), here rod-like structures have three axially-orientated hollow spaces. When the negative pressure of each hollow space is the same (Fig. 2.27 a, left), the rod is stressed with compression or buckling. Different negative pressures in the hollow spaces make the rod bend by different amounts (Fig. 2.27 a, right).

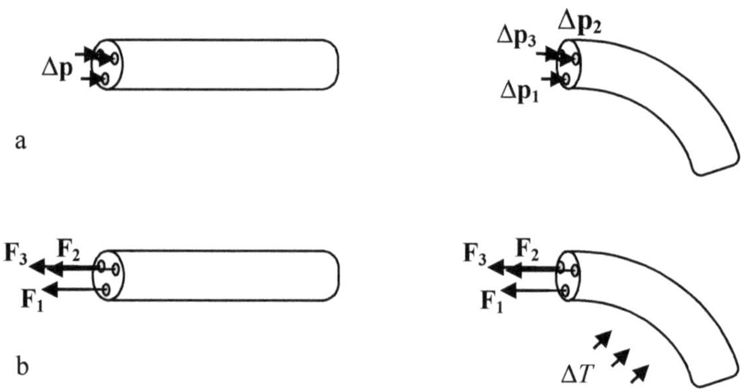

Fig. 2.27: Examples of using compliant system that can exhibit bifurcation behavior; a – a rod with three depressurized hollow spaces, left: the pressure in the hollow spaces have the same value where $\Delta p < 0$, right: the reduced pressure in one of the hollow spaces has a higher value where $|\Delta p_1| > |\Delta p_2|$, $\Delta p_2 = \Delta p_3$, thus the rod bends; b – a rod with three hollow spaces with threads made of a material such as a shape memory alloy, left: $F_1 = F_2 = F_3$, a subcritical state, right: a load $F_1 = F_2 = F_3$ is applied and the underside of the rod is warmed with ΔT, causing the rod to bend

An external load can be produced as in Fig. 2.27 b through the force of threads placed in the hollow spaces, made of a shape memory alloy, for example. When such wires are tensioned by the same amount (Fig. 2.27 b, left), where:

$$F_1 = F_2 = F_3 \,, \tag{2.10}$$

these stresses can be modeled by an axial force. This situation can thus be considered as a classic case of the stability problem (bifurcation) according to EULER with an axial force:

$$F = F_1 + F_2 + F_3 \,. \tag{2.11}$$

When various forces act on the rod, it bends accordingly.

Systems of this kind are compliant actuators, specifically fluid-mechanical actuators, when internal pressure is applied. Such actuators retain their shape, as long as they only experience a subcritical force. By using functional materials that change their compliance under the influence of external conditions (such as a temperature change ΔT) or conditions specifically altered by the user, a specific part of the system can be influenced either depending on its surroundings or the will of the user. If, for example, the temperature of the surroundings of the left actuator in Fig. 2.27 b rises, so that the material on its underside softens (due to its material properties), the system bends accordingly. Fig. 2.27 b (right) illustrates the case in question, where the temperature is increased and the three acting forces remain the same. Such a system can thus be designed so that it adapts to the conditions of its surroundings. Sensory and control processes are enormously simplified purely through using the mechanical properties of the material. The intelligently designed structural and geometric properties, as well as deliberately chosen materials, take on some of the sensing, information processing and transfer functions of the system.

3 Modeling compliant systems as rigid-body systems

One downside of compliant mechanisms is the complexity of describing them theoretically, especially given that analytical investigation is rarely possible. The analysis of compliant mechanisms can be simplified by establishing a *rigid-body model* to describe the relationship between the loads and displacements of compliant mechanisms. A rigid-body model is especially suitable for small deflections of compliant mechanisms. Models of this kind are described in [47], as well as other literature sources such as [52], [68] und [87]. In these cases, elastic elements are first identified and the loads acting upon them described, then the differential equations for each of the compliant elements are formulated and solved. Finally, a rigid-body model with a revolute joint and torsion spring is assembled for each elastic element. This quite laborious procedure can be avoided if a universally-applicable formula for the assembly of a rigid-body model can be derived. This chapter presents a formula of this kind in the form of mathematical equations, along with their derivations, which are finally combined into two general relationships between the geometrical and material parameters. These allow a rigid-body model of a compliant mechanism to be created easily and quickly.

3.1 Assumptions for modeling

The goal is to assemble a rigid-body model of a compliant system that contains compliant elements. It is assumed that the compliant mechanisms have purely elastic material properties and undergo only small deflections. Therefore, linear theory for the calculation of the deflections of elastic rods is used. For the classification of the compliance as either a concentrated or distributed compliance, the length of the whole system on the rod axis is taken as the maximum characteristic length of the system (see definition in Section 2.1.1).

Building the model described above is carried out in three steps. Initially, the compliant elements of the system are identified. These are areas of the system that exhibit a greatly increased compliance as compared to the remaining parts of the compliant mechanism. The compliant element connects two mechanism parts to one another, which are assumed to be rigid. Each compliant element is modeled as an elastic rod with constant cross-section fixed at one end. The other end remains free and is loaded with forces and moments. These loads are determined using the equilibrium conditions of the system parts for an established load situation of the system. In the second step, the compliant element is replaced with two rigid-body parts that are connected to one another with a pivot joint. The elasticity of the compliant element is therefore simulated as a torsion spring in the joint with a spring stiffness of c_t. In the third step, the rigid-body parts connected to one another by pivot joint with a spring and are introduced into the system and the relevance of this new rigid-body model is

https://doi.org/10.1515/9783110479744-039

checked, for example in its ability to move by means of the *degree of freedom* it exhibits. Fig. 3.1 shows the three steps described above using the example of a pair of compliant pliers. The first and final steps are specific to the system and depend on the structural and geometric design of the given system, as well as on the loads acting upon it. Therefore, these steps should be carried out differently for each individual system. In contrast, the second step can be generalized for all systems, as described below.

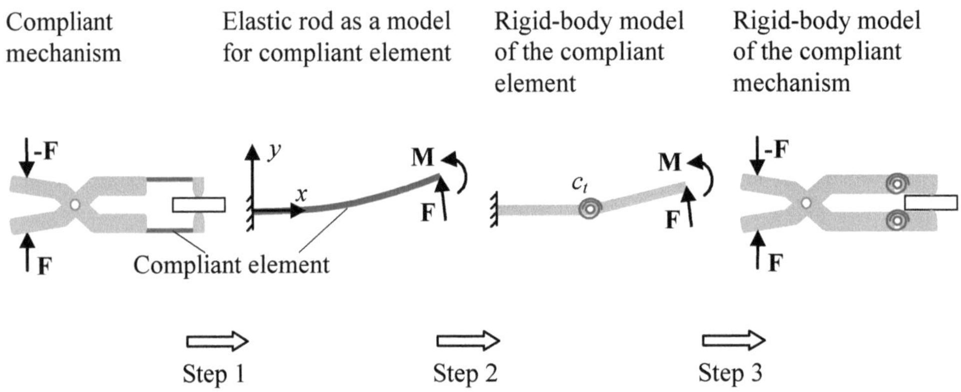

Fig. 3.1: A schematic of the modeling steps for the transformation of a compliant system (partially compliant mechanism, see Table 2.1) into a rigid-body model using the example of a pair of pliers

In the second modeling step, a compliant element is replaced with two rigid-body parts that are joined together. The total length of both rigid-body parts and the length of the compliant element are identical and are denoted with l (Fig. 3.2 a-b). The relationship of a moving rigid-body part to the length l is described with the parameter δ (see [47]). Thus δl is the length from the joint to the free end of a rigid-body part and $l(1-\delta)$ is the distance from the fixed end to the joint. The parameter δ therefore also describes the position of the joint along the length l in the rigid-body model. A torsion spring with a spring constant c_t is attached between the two rigid-body parts. The deflections of the compliant element are assumed to be very small. Therefore, the angle θ of the relative rotation between the two rigid-body parts is also assumed to be small. The rigid-body parts joined together in this way, along with the torsion spring, thus form a rigid-body model of a compliant element. The two parameters δ and c_t should be determined, in order to create this rigid-body model of a compliant element.

A compliant element which is treated as an elastic rod and which is fixed at one end, is loaded with different forces and moments. These loads are broken down into three groups:

- $n = 1$: load due to a moment,
- $n = 2$: load due to a force,
- $n = 3$: load due to a distributed transverse force.

The parameter n describes the type of load and allows recursive notation to be used. The designations of the loads and the inclination angles of the compliant element are given with the index “0”, although this does not apply to the parameters of the rigid-body model. The three loads listed above are now considered separately, resulting in three separate cases.

Case 1: Load due to a moment

An elastic rod with length l is fixed at one end and loaded with a moment $\mathbf{M_0}$ at its other free end (Fig. 3.2 a). The displacement of a point on the rod with coordinate x in the y-direction is described as $u_y(x)$.

$$u_y''(x) = \frac{M_0}{EI_z} \tag{3.1}$$

By using the following boundary conditions:

$$\begin{aligned} u_y(0) &= 0, \\ u_y'(0) &= 0, \end{aligned} \tag{3.2}$$

the inclination of the tangent $u_y'(l)$ at the end of the elastic rod to the x-axis and the corresponding displacement $u_y(l)$ is found:

$$\begin{aligned} u_y'(l) &= \frac{M_0 l}{EI_z}, \\ u_y(l) &= \frac{M_0 l^2}{2EI_z}. \end{aligned} \tag{3.3}$$

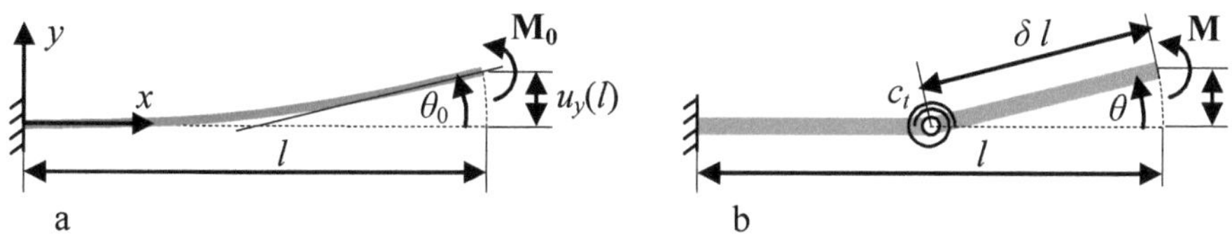

Fig. 3.2: Load due to a pure moment; a – elastic rod; b – rigid-body model loaded with a moment

The following applies for the angle of inclination of the free rigid-body part in Fig. 3.2 b:

$$\theta = \frac{M}{c_t}. \tag{3.4}$$

Case 2: Load due to a force

The differential equation in the case of a load due to a force $\mathbf{F_0}$ acting on the free end of the elastic rod (Fig. 3.3 a) is written as:

$$u_y''(x) = \frac{F_0(l-x)}{EI_z}. \tag{3.5}$$

Taking the boundary conditions in (3.2) into account gives the inclination of the tangent and the displacement of the rod at the point $x = l$:

$$\begin{aligned} u_y'(l) &= \frac{F_0 l^2}{2EI_z}, \\ u_y(l) &= \frac{F_0 l^3}{3EI_z}. \end{aligned} \tag{3.6}$$

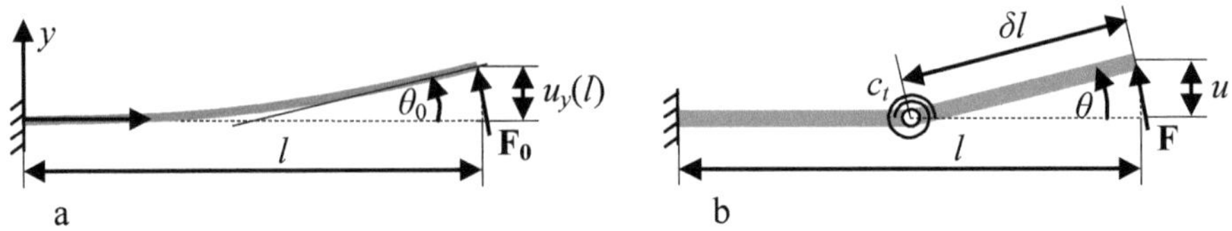

Fig. 3.3: Load due to a force; a – elastic rod; b – rigid-body model in the case of a load due to a force

To find the angle of inclination of the free rigid-body part in Fig. 3.3 b, the moment of the force about the pivot joint is taken into account.

$$\theta = \frac{F\delta l}{c_t} \tag{3.7}$$

Case 3: Load due to a distributed force

The following applies for the elastic rod in Fig. 3.4 a, which is loaded with a distributed force:

$$u_y''(x) = \frac{q_0 (l-x)^2}{2EI_z}. \tag{3.8}$$

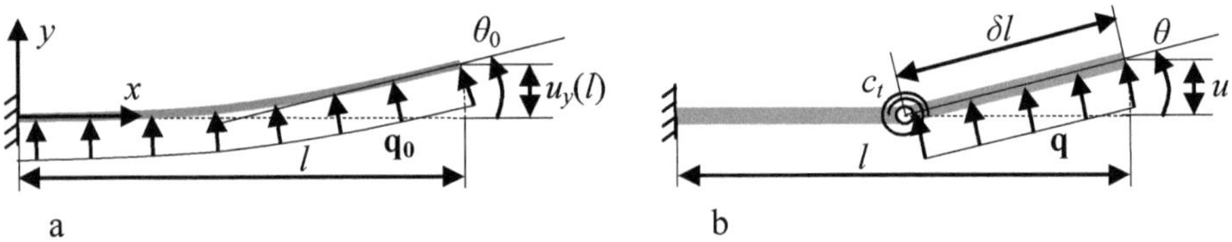

Fig. 3.4: Load due to a distributed force; a – elastic rod; b – rigid-body model in the case of a load due to a distributed force

Taking into account the boundary conditions in (3.2), the inclination of the tangent and the displacement for $x = l$ can be written as follows:

$$u_y'(l) = \frac{q_0 l^3}{6EI_z}, \qquad u_y(l) = \frac{q_0 l^4}{8EI_z}. \tag{3.9}$$

The angle of inclination of the rigid-body part is described using the distributed load (Fig. 3.4 b), which only acts on the line δl.

$$\theta = \frac{q(\delta l)^2}{2c_t} \tag{3.10}$$

The parameter δ and the spring constant c_t should be found for each of these three cases.

3.2 Modeling for individual load cases

Auxiliary moments with the following designations are introduced, allowing all three cases to be considered generally. For this elastic rod, these are as follows:

$$
\begin{aligned}
M_{01}{}^{H} &= M_{0},\\
M_{02}{}^{H} &= F_{0}l, \qquad (3.11)\\
M_{03}{}^{H} &= \frac{1}{2}q_{0}l^{2}.
\end{aligned}
$$

These auxiliary moments are also internal moments at the fixed end of the rod for each load case. Three auxiliary moments are also introduced for the rigid-body model:

$$
\begin{aligned}
M_{1}{}^{H} &= M,\\
M_{2}{}^{H} &= Fl, \qquad (3.12)\\
M_{3}{}^{H} &= \frac{1}{2}ql^{2}.
\end{aligned}
$$

Using these auxiliary moments, general expressions for the inclinations of the tangent $u_y'(l)$ and the displacement of the end of the elastic rod $u_y(l)$ for each load case (equations (3.3), (3.6) and (3.9)) are found:

$$
\begin{aligned}
u_y'(l) &= \frac{l}{EI_z}\frac{M_{0n}{}^{H}}{n},\\
u_y(l) &= \frac{l^2}{EI_z}\frac{M_{0n}{}^{H}}{n+1}. \qquad (3.13)
\end{aligned}
$$

The angle of inclination of the moving rigid-body part θ, as well as its displacement u in the y-direction are written for each case as follows:

$$
\begin{aligned}
\theta &= \frac{M_n{}^{H}\delta^{n-1}}{c_t},\\
u &= \delta l \theta, \qquad (3.14)
\end{aligned}
$$

assuming the deflections are small, therefore it can be applied:

$$\sin\theta \approx \theta . \qquad (3.15)$$

This last expression in equation (3.14) is purely geometric in character, and depends neither on the loads nor on the load case n. The derivative of the displacement of the end of the elastic rod $u_y'(l)$ is the tangent of the angle θ_0 for each load case n, and can be considered to be approximately equal to the angle θ_0, because this angle is very small:

$$u_y'(l) = \tan\theta_0 \approx \theta_0 . \qquad (3.16)$$

Two models of a compliant element, an elastic rod and a rigid-body, are considered equivalent when the following three conditions are met. The angles of inclination of both systems θ_0 and θ should be identical. Since these angles are very small, this condition can be written as follows:

$$u_y'(l) = \theta . \tag{3.17}$$

Furthermore, the displacements of the end of the rod $u_y(l)$ and the end of the moving rigid-body part u should be the same:

$$u_y(l) = u . \tag{3.18}$$

Finally, the loads acting on the elastic rod and the rigid-body part for each n ($n = 1,2,3$) should be identical:

$$M_n^H = M_{0n}^H . \tag{3.19}$$

Considering conditions (3.17)-(3.19) and by using the equations (3.13) and (3.14), the following equations for the position of the joint and the spring constant in each load case n are determined:

$$\delta = \frac{n}{n+1}, \tag{3.20}$$

$$c_t = \frac{n^n}{(n+1)^{n-1}} \frac{EI_z}{l} . \tag{3.21}$$

Table 3.1: Values for the positions of the joints and corresponding spring constants of a rigid-body model in individual load cases

n	δ	c_t	Depiction
1	$\frac{1}{2}$	$\frac{EI_z}{l}$	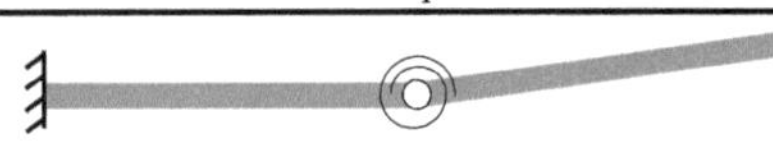
2	$\frac{2}{3}$	$\frac{4}{3}\frac{EI_z}{l}$	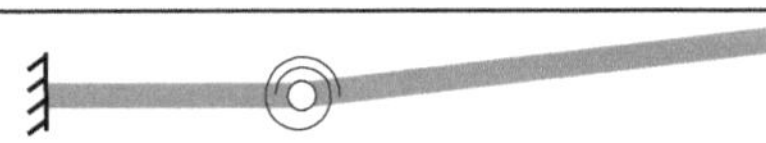
3	$\frac{3}{4}$	$\frac{27}{16}\frac{EI_z}{l}$	

In line with the equations (3.20) and (3.21), the position of the joints and the corresponding spring constants in individual load cases for the rigid-body models can be found. This provides the following values for the different cases n, which are summarized in Table 3.1. According to these results, the joint is "displaced" from the center of the system in the direction of the fixed end with the rising the parameter n. Meanwhile, the stiffness of the spring is increased. It should also be noted that the results for the spring constant and the position of the joint in the rigid-body model depend only on the load case n, and not on the magnitude of the load.

3.3 Modeling for complex loads

Complex loads often act on a compliant element in a compliant mechanism; this element is modeled as an elastic rod. This rod is simultaneously loaded with a force, moment and distributed force. Since linear theory is used here, the superposition principle applies. As a result, displacements and angles of the rod for individual load cases can be summed together to find the corresponding parameters for a complex load:

$$u_y'(l) = \frac{l}{EI_z} \sum_{n=1}^{3} \frac{M_{0n}{}^H}{n}, \tag{3.22}$$

$$u_y(l) = \frac{l^2}{EI_z} \sum_{n=1}^{3} \frac{M_{0n}{}^H}{n+1}. \tag{3.23}$$

Similarly, the following relationships for a rigid-body system apply:

$$\theta = \frac{\sum_{n=1}^{3} M_n{}^H \delta^{n-1}}{c_t}, \tag{3.24}$$

$$u = \delta l \theta. \tag{3.25}$$

The equation (3.25) includes only geometrical dependencies between the parameters. When establishing an equivalent rigid-body model of a compliant element under a complex load, conditions (3.17)-(3.19) also apply. By using these conditions, and taking equations (3.22)-(3.25) into account, the position of the joint and the value of the spring constant for a rigid-body model can be found:

$$\delta = \frac{\sum_{n=1}^{3} \frac{M_n{}^H}{n+1}}{\sum_{n=1}^{3} \frac{M_n{}^H}{n}}, \tag{3.26}$$

$$c_t = \frac{EI_z}{l} \frac{\sum_{n=1}^{3} M_n{}^H \delta^{n-1}}{\sum_{n=1}^{3} \frac{M_n{}^H}{n}}. \tag{3.27}$$

The equations (3.26) and (3.27) are general relationships used to find the position of the joint and the spring stiffness for a rigid-body model.

Using the two equations (3.26) and (3.27), possible values of parameters δ and c_t can be generally investigated. To this end, equation (3.26) will first be rewritten as follows:

$$\delta = \frac{\sum_{n=1}^{3} \frac{M_n{}^H}{n+1}}{\sum_{n=1}^{3} \frac{M_n{}^H}{n+1} + \sum_{n=1}^{3} \frac{M_n{}^H}{n(n+1)}}. \tag{3.28}$$

When the two sums of the denominator in the equation (3.28) have the same sign, it is clear that the joint lies within the length of the rod. For parameter δ, this means that:

$$\delta \in [0,1]. \tag{3.29}$$

To give a specific example, two types of load can be considered in the system. Additionally, the auxiliary moments have the same sign. In this case, the joint of the rigid-body model lies in the range between the values of parameter δ that correspond to the individual load cases (Table 3.1). This is also true for the value of the spring stiffness. If, for example, a moment ($n = 1$) and a force ($n = 2$) act on a compliant system, then the following applies for the position of the joint:

$$\delta = \frac{3M_1^H + 2M_2^H}{6M_1^H + 3M_2^H}. \tag{3.30}$$

If a parameter k_M as a quotient for the parameters M_1^H and M_2^H is introduced, then the following applies:

$$\delta = \frac{3k_M + 2}{6k_M + 3} \quad \text{with } k_M = \frac{M_1^H}{M_2^H}. \tag{3.31}$$

In the range $k_M \in [0,\infty]$, which corresponds to the same sign of both auxiliary moments, δ decreases monotonously as parameter k_M increases. The following boundary values apply for δ from (3.31):

$$\lim_{k_M \to 0} \delta = \frac{2}{3}, \qquad \lim_{k_M \to \infty} \delta = \frac{1}{2}. \tag{3.32}$$

Thus, the position of the joint in the rigid-body model for a complex load due to a moment and a force with the same sign corresponds to the range of δ:

$$\delta \in \left[\frac{1}{2}, \frac{2}{3}\right]. \tag{3.33}$$

The boundary values of this range correspond to the cases of the individual loads: the load due to the moment as in $n = 1$ and the load due to the force as in $n = 2$ (compare with Table 3.1). Similarly, it can be shown for the complex loads mentioned above, that the spring stiffness of the torsion spring lies within the following range:

$$c_t \in \left[\frac{EI_z}{l}, \frac{4}{3}\frac{EI_z}{l}\right]. \tag{3.34}$$

Therefore, a range can be given for both parameters δ and c_t for specific loads, within which these parameters can be expected to lie, without needing to perform the calculations required.

3.4 Modeling for concentrated compliance

Due to the relatively small lengths of compliant elements with concentrated compliance, the corresponding revolute joint in the rigid-body model can be placed at the middle point of length l. Additionally, only two types of load are considered: a load due to a moment, and a load due to a force. Due to the small length of the rod, the distributed force can be disregarded. If the general equation from (3.27) is used to calculate the spring constant while taking the following assumptions into account:

$$M_{03} = M_3 = 0,$$
$$\delta = \frac{1}{2} \tag{3.35}$$

a simple expression for the spring constant is attained, which is neither dependent on the load case nor on the complexity of the load:

$$c_t = \frac{EI_z}{l}. \tag{3.36}$$

3.5 Comparison of the methods

The question remains of how well the results attained when using a rigid-body model of this kind to describe deformations and loads compare to those of other models. In order to discuss this question, a comparison will be made of the deflections of a compliant rod according to the rigid-body model, according to linear theory and according to non-linear theory. Non-linear theory, which allows for large deflection, will be introduced later in Chapter 4, but selected equations from (4.148) and (4.152) will be brought forward in order to make this comparison. As an example used to validate this model, an elastic rod fixed at one end under the effect of a moment, a force and a distributed force is considered. Fig. 3.5 shown the results for the deflections in the x- and y-directions for the load case $n = 2$, using the three methods mentioned above.

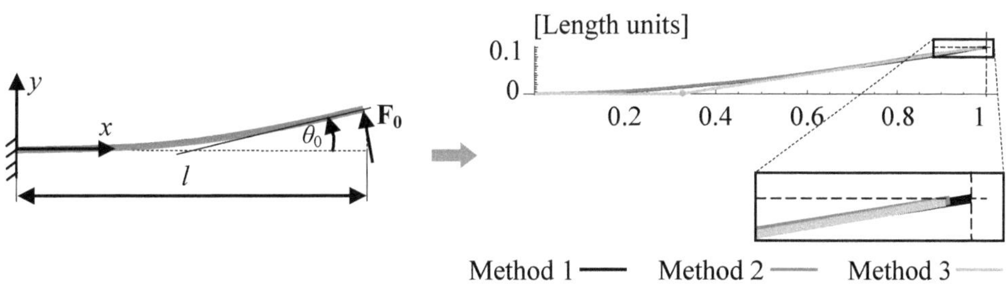

Fig. 3.5: A comparison of the deflections of the free rod end due to a force ($n = 2$) using three different modeling methods; 1: linear theory, 2: non-linear theory, 3: rigid-body model

The displacements in the x- and y-directions are given as u_x and u_y. Equilibrium conditions for a rod element of length ds are formulated for the rod's deformed position, in order to derive the equations describing the deflections according to non-linear theory (see Chapter 4). The following designations are used in these equations: Q_i with $i = 1{,}2$ are the internal forces acting on the rod, where Q_1 is a longitudinal force and Q_2 is a shear force; M_3 is a bending moment and κ_3 is the curvature of the axial line of the rod. *Axial line* of the rod runs through the centroids of its cross-sections. A moment, a force and a distributed force are chosen, which lead to a displacement u_y of 10 % of the rod-length according to Method 1. The results for the displacements in two directions for each load case are given in Table 3.2, broken down by the three methods used.

Table 3.2: Comparison of the results for the displacement at the rod's end using three different methods: linear theory, non-linear theory and using the rigid-body model; the three loads were chosen that would cause a displacement of 10 % of the rod's length at its end according to method 1, this length is given as length 1 (in length units)

Load case and its value	Displacement	Method 1: linear theory		Method 2: non-linear theory (large deflection)		Method 3: rigid-body model	
		Equations	Results	Equations	Results [length units]	Equations	Results [length units]
$n = 1$ $\frac{M_0}{EI_z} = 0.2$ [length units]	u_x	$u_x = 0$ $u_y = \frac{l^2}{EI_z}\frac{M_n^H}{n+1}$	0	$\frac{dQ_1}{ds} - Q_2\kappa_3 = 0$ $\frac{dQ_2}{ds} + Q_1 + q_2 = 0$ $\frac{dM_3}{ds} + Q_2 = 0$ $\frac{d\theta_3}{ds} = \frac{M_3}{EI_z}$ $\frac{du_x}{ds} = \cos\theta_3 - 1$ $\frac{du_y}{ds} = \sin\theta_3$	–0.00665	$u_x = \delta l\cos\theta - \delta l$ $u_y = \delta l\,\theta$	–0.00996
	u_y		0.1		0.09966		0.1
$n = 2$ $\frac{F_0}{EI_z} = 0.3$ [length units^{-1}]	u_x		0		–0.00599		–0.00748
	u_y		0.1		0.09974		0.1
$n = 3$ $\frac{q_0}{EI_z} = 0.8$ [length units^{-2}]	u_x		0		–0.00570		–0.00665
	u_y		0.1		0.09977		0.1

The differences in the solutions when using linear theory, non-linear theory and the rigid-body model are smaller than one percent. If such a discrepancy is acceptable, the rigid-body model can be used for displacements smaller than 10 % of the length of the rod. It should be noted that the displacement in the *x*-direction using the rigid-body model is not zero. Thus, $u_x(l)$ in the rigid-body model is closer in value to the solution provided by non-linear theory.

3.6 Serial cascading rigid-body joints

The *degree of freedom* of the achieved rigid-body model of a compliant mechanism is to be investigated. The mechanism found in this model should firstly be able to move, with a degree of freedom of at least one, while also exhibiting a simple structure. If this is not the case, the degree of freedom of the rigid-body system can be adjusted by reducing or increasing the number of joints in the system. Two or more joints of the rigid-body system can be replaced with a single joint. This reduces the degree of freedom of the system and simplifies the model. Conversely, the degree of freedom increases as the number of joints increases.

Using the example of the double-jointed rigid-body system in Fig. 3.6 a, replacing two joints with a single joint (Fig. 3.6 b) is discussed. The designations are taken from Fig. 3.6 and both systems should have the same total length:

$$l_1 + l_2 + l_3 = l_0 + l \,. \tag{3.37}$$

The following applies for both of the joints in the system shown in Fig. 3.6 a:

$$\theta_1 c_{t1} = M + Fl_1,$$
$$\theta_2 c_{t2} = M + F(l_1 + l_2). \tag{3.38}$$

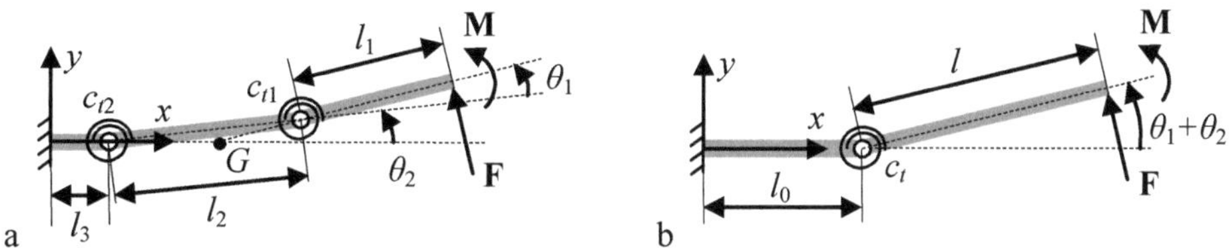

Fig. 3.6: Two rigid-body systems; a – a system with two joints; b – a system with one joint

Multiplying the first equation from (3.38) with c_{t2}, the second equation with c_{t1} and finally adding both equations together results in the following expression:

$$(\theta_1 + \theta_2)c_{t1}c_{t2} = c_{t1}\left(M + F(l_1 + l_2)\right) + c_{t2}(M + Fl_1). \tag{3.39}$$

The equation in (3.39) is now rewritten simply by rearranging the terms and adding the expression $(M + Fl)$:

$$(\theta_1 + \theta_2)\frac{c_{t1}c_{t2}(M + Fl)}{c_{t1}\left(M + F(l_1 + l_2)\right) + c_{t2}(M + Fl_1)} = M + Fl. \tag{3.40}$$

The right-hand side of the expression in (3.40) forms a moment that acts on the rigid-body system in Fig. 3.6 b. This moment is given according to the joints. The left-hand side of the expression in (3.40) gives the product of the total angle $\theta_1 + \theta_2$ and the spring constant c_t for the system in Fig. 3.6 b. Thus, the spring constant for this system is:

$$c_t = \frac{c_{t1}c_{t2}(M + Fl)}{c_{t1}\left(M + F(l_1 + l_2)\right) + c_{t2}(M + Fl_1)}. \tag{3.41}$$

The spring constant c_t of the new system (Fig. 3.6 b) can also be given in the following form:

$$\frac{1}{c_t} = \frac{1}{c_{t2}} \cdot \frac{M + F(l_1 + l_2)}{M + Fl} + \frac{1}{c_{t1}} \cdot \frac{M + Fl_1}{M + Fl}. \tag{3.42}$$

The position of this joint is now investigated. This is shown as point G in Fig. 3.6 and is dependent on the length l of the moving rigid-body part in Fig. 3.6 b. The angle of inclination of both systems (Fig. 3.6 a and b) are the same, since the two systems should be equivalent:

$$l(\theta_1 + \theta_2) = l_2\theta_2 + l_1(\theta_1 + \theta_2). \tag{3.43}$$

According to (3.43), the following applies for l:

$$l = \frac{l_2\theta_2 + l_1(\theta_1 + \theta_2)}{\theta_1 + \theta_2}. \tag{3.44}$$

The angles θ_1 and θ_2 are written as follows, according to the equations in (3.38):

$$\begin{aligned} \theta_l &= \frac{M + Fl_1}{c_{t1}}, \\ \theta_2 &= \frac{M + F(l_1 + l_2)}{c_{t2}} \end{aligned} \tag{3.45}$$

and then considering equation (3.44):

$$l = \frac{M\left(c_{t1}(l_1 + l_2) + c_{t2}l_1\right) + F\left(c_{t1}(l_1 + l_2)^2 + c_{t2}l_1^{\,2}\right)}{M\left(c_{t1} + c_{t2}\right) + F\left(c_{t1}(l_1 + l_2) + c_{t2}l_1\right)}. \tag{3.46}$$

The expressions in (3.42) and (3.46) determine the spring constant and the position of the joint for the system in Fig. 3.6 b, which is equivalent to the original system with two joints in Fig. 3.6 a. Two equivalent systems are defined here as systems that undergo an identical displacement in the y-direction and have the same inclination angle when loaded with identical loads.

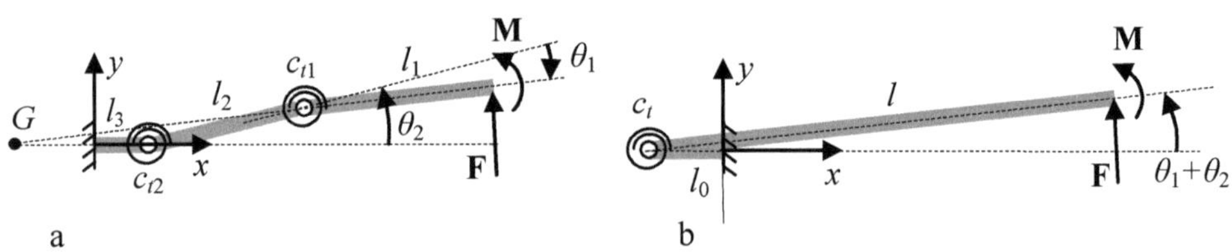

Fig. 3.7: Depiction of a system with a negative angle θ_1 and a positive angle θ_2; a – a system with two joints; b – realization of a single joint to the left of the fixing point, l_0 is negative

A negative angle θ_1 results in a virtual joint, as shown by point G in Fig. 3.7 a. However, a joint of this kind is only possible if it is added to the left of the clamping point, as can be seen in Fig. 3.7 b.

Generally, the parameters c_t and l depend on the loads M and F, as in (3.42) and (3.46). The position of the joint and the stiffness of its spring also change with varying loads.

Table 3.3 lists a general case, as well as three other cases. If only a single moment or force acts on a rigid-body system, the expressions for the spring constant and the rigid-body length l do not depend on the moment or force. This also applies when the moment M depends on the force F and can be written as a product of h (i.e. a lever arm) and F. If only a moment acts on the system ($M \neq 0$, $F = 0$), the expression for the spring constant is equal to that of serially-arranged tension springs. Other situations, e.g. for $l_1 = l_2$ or $c_{t1} = c_{t2}$, can be modeled according (3.42) and (3.46).

Table 3.3: Relationship between the parameters of the rigid-body system with a joint from Fig. 3.6 b and the rigid-body system with two joints from Fig. 3.6 a; the systems are equivalent

Case	c_t	l
$M \neq 0$, $F \neq 0$, $l_1 \neq l_2$, $c_{t1} \neq c_{t2}$	$\frac{1}{c_t}=\frac{1}{c_{t2}}\cdot\frac{M+F(l_1+l_2)}{M+Fl}+\frac{1}{c_{t1}}\cdot\frac{M+Fl_1}{M+Fl}$	$l=\frac{M\left(c_{t1}(l_1+l_2)+c_{t2}l_1\right)+F\left(c_{t1}(l_1+l_2)^2+c_{t2}l_1^2\right)}{M\left(c_{t1}+c_{t2}\right)+F\left(c_{t1}(l_1+l_2)+c_{t2}l_1\right)}$
$M \neq 0$, $F = 0$, $l_1 \neq l_2$, $c_{t1} \neq c_{t2}$	$\frac{1}{c_t}=\frac{1}{c_{t2}}+\frac{1}{c_{t1}}$	$l=\frac{c_{t1}(l_1+l_2)+c_{t2}l_1}{c_{t1}+c_{t2}}$
$M = 0$, $F \neq 0$, $l_1 \neq l_2$, $c_{t1} \neq c_{t2}$	$\frac{1}{c_t}=\frac{1}{c_{t2}}\cdot\frac{l_1+l_2}{l}+\frac{1}{c_{t1}}\cdot\frac{l_1}{l}$	$l=\frac{c_{t1}(l_1+l_2)^2+c_{t2}l_1^2}{c_{t1}(l_1+l_2)+c_{t2}l_1}$
$M = hF$, $F \neq 0$, $l_1 \neq l_2$, $c_{t1} \neq c_{t2}$	$\frac{1}{c_t}=\frac{1}{c_{t2}}\cdot\frac{h+l_1+l_2}{h+l}+\frac{1}{c_{t1}}\cdot\frac{h+l_1}{h+l}$	$l=\frac{c_{t1}(l_1+l_2)(h+l_1+l_2)+c_{t2}l_1(h+l_1)}{c_{t1}(h+l_1+l_2)+c_{t2}l_1}$

3.7 Modeling examples

This section will use three examples to demonstrate the application of the method of modeling a compliant mechanism with a rigid-body model. The first example deals with a simple gripping system with two compliant joints. Two further examples will then be presented, in order to show that some systems are not well suited to the introduction of a single joint to replace a compliant element in a rigid-body model. This leads to the degree of freedom of a rigid-body system with a value of zero, as shown in the second example. The third example presents a rigid-body model of a compliant mechanism, where the use of the modeling method presented here results in an infinitely large parameter δ. On the basis of these specific cases, problems that might arise when using this modeling method are highlighted and solutions are suggested. Since solutions are provided in general form for all examples, these results can be applied to similar systems.

3.7.1 A gripping system with two joints

A compliant gripping system (Fig. 3.8 a) is used to grip an object, which is held by a contracting actuating element. The compliant finger undergoes a deformation due to a gripping force $\mathbf{F}_{Gr}$. In this section, a rigid-body model is used to simulate this example. The length L_0 of the compliant joint is assumed to be much smaller than the length of the compliant gripper in the same direction. The length L is comparable to the length of the whole system. Due to the geometrical parameters, the element of the length L_0 is seen as a compliant joint with concentrated compliance, while the element of the length L exhibits distributed compliance. In the case of the compliant joint with concentrated compliance, a pivot joint is positioned in the middle of the length L_0 of the rigid-body model. A torsion spring with a stiffness c_{tL0} is connected to this joint.

$$\delta_{L0}=\frac{1}{2},\qquad c_{tL0}=\frac{EI_z}{L_0} \tag{3.47}$$

The bending stiffness of the two compliance gripping elements is denoted with EI_z. The position of the joint and the spring stiffness of the torsion spring for the rigid-body model should be found for the finger with length L, which exhibits distributed compliance and is loaded with a gripping force. This is a pure load case $n = 2$; the values for δ_L and c_{tL} are taken from Table 3.1:

$$\delta_L = \frac{2}{3}, \qquad c_{tL} = \frac{4}{3}\frac{EI_z}{L}. \tag{3.48}$$

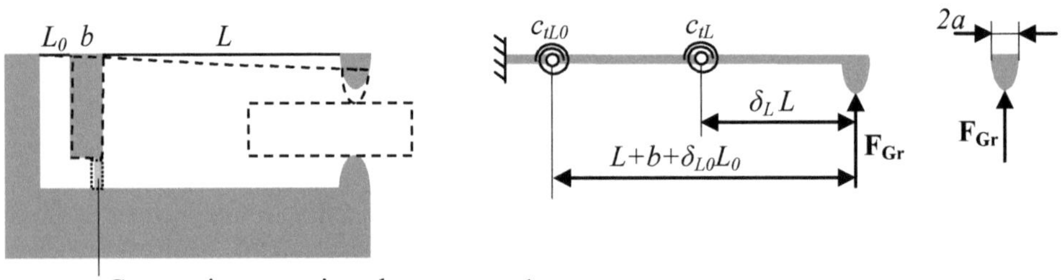

Fig. 3.8: Example of a compliant gripping system; a – a gripping system with a compliant joint exhibiting concentrated compliance and a joint with the distributed compliance; b – a corresponding rigid-body model; c – gripper jaw

This results in a rigid-body system for this gripping system (Fig. 3.8 b). If the dimension of the gripping jaw, denoted here as $2a$, should be taken into account (Fig. 3.8 c), then an additional moment also acts on the compliant element with the length L. The auxiliary moments are introduced as follows:

$$M_1^H = F_{Gr}a, \qquad M_2^H = F_{Gr}L, \qquad M_3^H = 0. \tag{3.49}$$

The position of the joint and the stiffness of the corresponding torsion spring are found according to (3.26) and (3.27):

$$\delta_L = \frac{3a+2L}{6a+3L}, \qquad c_{tL} = \frac{EI_z}{L}\frac{2a+2L\delta_L}{2a+L}. \tag{3.50}$$

Both of these parameters are independent of the value of the loads, but do depend on the geometric dimensions and the material properties of the compliant mechanism. Fig. 3.8 b shows a rigid-body model of this compliant gripping system. The rigid-body model can be used to establish the relationship between the displacements and loads, as well as to simulate the deformation of the compliant gripping system. It should be noted that this model is only suitable for small deflections.

3.7.2 A gripping system with multiple joints

A gripping system with four compliant elements, organized into two pairs of lengths L and L_0 (Fig. 3.9 a). A rigid-body system is used to model the opening of the gripping fingers due to an actuating force $\mathbf{F_{An}}$. Based on geometric dimensions, the compliant element with length L is modeled as a compliant joint with distributed compliance and the compliant element with length L_0 as a joint with concentrated compliance. Firstly, the loads acting on the compliant elements should be found. To this end, the following conditions are taken into consideration. Due to the forces and moments acting upon it, the gripping finger is in a state of equilibrium (Fig. 3.9 b). The axial forces acting on the compliant elements are disregarded. Additionally, the two tangents t_1 and t_2 of the compliant elements remain parallel at the junction of the rigid-body. The following relationships for the acting forces and moments are taken from the first condition, while introducing the actuating force:

$$\begin{aligned} F_L &= F_{L0} = \frac{1}{2} F_{An}, \\ M_L - M_{L0} &= \frac{F_{An}}{2} a. \end{aligned} \tag{3.51}$$

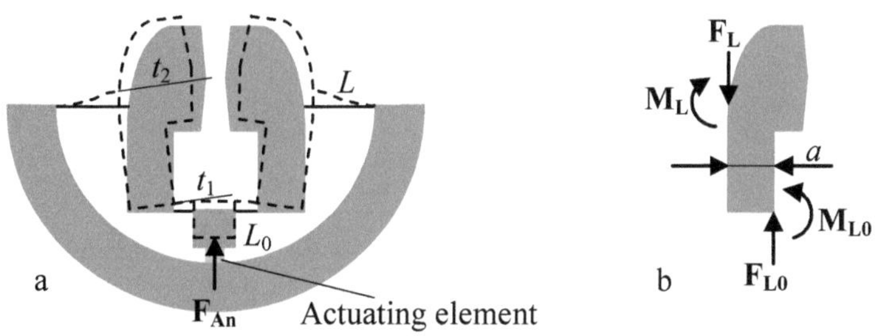

Fig. 3.9: Example of a compliant gripping system with multiple joints; a – a gripping system with two joints exhibiting distributed compliance and two joints exhibiting concentrated compliance; b – representation of the loads acting on a gripping element (in equilibrium)

The expression from (3.22) is used for the tangential gradients. Taking into account the parallelism of the tangents t_1 and t_2, the moments M_{L0} and M_L acting on the compliant elements with corresponding lengths L_0 and L can be found:

$$M_{L0} = \frac{F_{An}}{2} h, \quad M_L = \frac{F_{An}}{2} h. \tag{3.52}$$

The following applies for the parameter h in this example:

$$h = \frac{L_0^{\,2} - 2aL_0 - L^2}{2(L - L_0)}. \tag{3.53}$$

The joint of the rigid-body model replacing the compliant elements with concentrated compliance is positioned in the middle of the length L_0. The spring constant can be found using (3.36). A force and a moment with values F_L and M_L act on the compliant element with length L in the opposite direction to the vectors $\mathbf{F_L}$ and $\mathbf{M_L}$ from Fig. 3.9 b:

$$\begin{aligned} M_1^H &= \frac{F_{An}}{2} h, \\ M_2^H &= \frac{F_{An}}{2} L, \\ M_3^H &= 0. \end{aligned} \tag{3.54}$$

The position of the joint and the spring stiffness can be found using (3.26) and (3.27). It should be noted that the parameter h is negative, since $L > L_0$. Therefore, the position of the joint is not guaranteed to lie within the length of the compliant element for all values of parameters L, L_0 and a. (see also Section 3.3). Fig. 3.10 b shows a rigid-body system for the gripping system, where $0 < \delta_L < 1$. The following expressions apply for the position of the joint and the spring constant:

$$\begin{aligned} \delta_L &= \frac{3h + 2L}{6h + 3L}, \\ c_{tL} &= \frac{EI_z}{L} \frac{2h + 2L\delta_L}{2h + L}. \end{aligned} \tag{3.55}$$

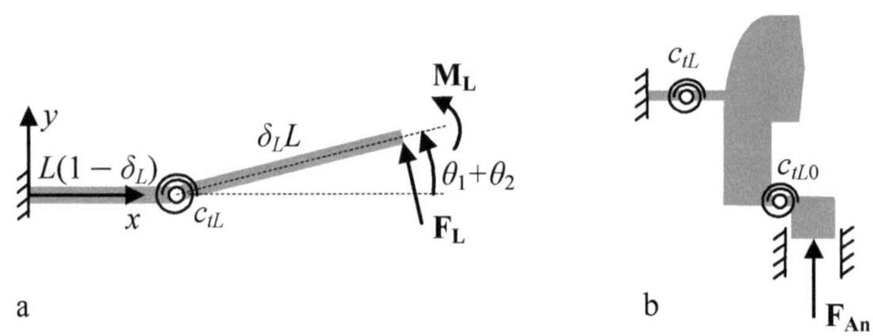

Fig. 3.10: A rigid-body model; a – a rigid-body model of a compliant element with length L; b – a rigid-body model of the gripping system with a degree of freedom of zero

The degree of freedom of the resulting system in Fig. 3.10 b is zero, meaning no relative movement of the gripping body can take place. The degree of freedom must be increased to allow the system to move. This is possible by introducing an additional joint into the system, resulting in a model of rigid-body with length L and two joints (Fig. 3.11 a). Since the force F_L and the moment M_L are mutually dependent, as is clear from (3.51) and (3.52), the last row of Table 3.3 shows the relationship between the parameters of the single-jointed system in Fig. 3.10 a and of the new double-jointed rigid-body systems in Fig. 3.11 a. The positions of the joints, as well as their spring stiffnesses, remain independent of the loads F_L and M_L.

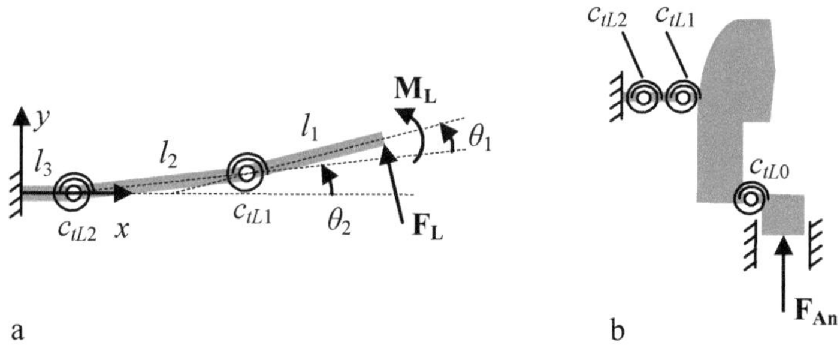

Fig. 3.11: A rigid-body model; a – a rigid-body model with two joints for a compliant element of length L; b – a rigid-body model of the gripping system with a degree of freedom of one

Two equations (in the last row of Table 3.3) can be used to find the parameters of the two new joints. The following assumptions are made, in order to solve these equations:

$$\begin{aligned} c_{tL2} &= c_{tL1}, \\ l_2 &= l_1. \end{aligned} \tag{3.56}$$

Applying the relevant parameters to the equations from the last row of Table 3.3 and taking the assumptions from (3.56) into account results in the following expressions:

$$\begin{aligned} \delta_L L &= l_1 \frac{3h+5l_1}{h+3l_1}, \\ \frac{1}{c_{tL}} &= \frac{1}{c_{tL1}} \frac{2h+3l_1}{h+\delta_L L}. \end{aligned} \tag{3.57}$$

The first equation from (3.57) is used to determine l_1, and c_{tL1} can be found using the second equation. It should be noted that the parameter h can be negative in the case of $L > L_0$. Therefore, the relevance of this solution should be verified. These parameters determine the new rigid-body model (Fig. 3.11 b) for the system from Fig. 3.9 a.

3.7.3 Parallel cascading compliant elements

A mechanical compliant system of an optical sensor unit (Fig. 3.12) can be used as a further example. The compliant system consists of two compliant elements arranged parallel to one another, which form an elastic connection between a frame and a moving part of the system. When a force $2\mathbf{F_L}$ acts on the system, this moving system element undergoes a displacement (Fig. 3.13 a-b). This displacement is assumed to be approximate to a translational movement. A rigid-body model should be created for this compliant system. One of the two compliant elements is considered and modeled as an elastic rod. This rod is fixed at one end where $x = 0$, and the other end is loaded with a force F_L and a moment M_L (Fig. 3.13 a-b). Firstly, auxiliary moments are introduced:

$$\begin{aligned} M_1^H &= M_L, \\ M_2^H &= -F_L L, \\ M_3^H &= 0. \end{aligned} \tag{3.58}$$

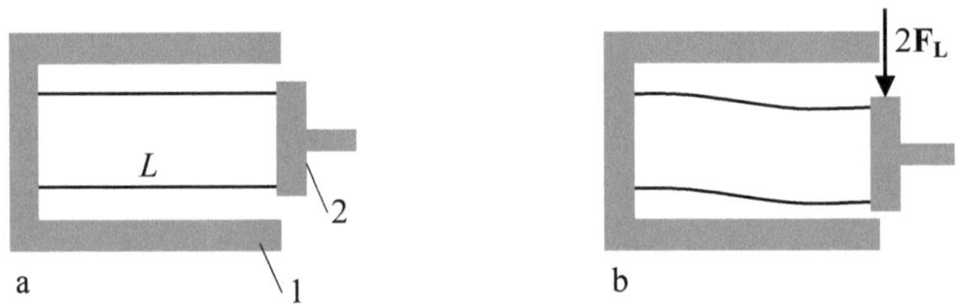

Fig. 3.12: Mechanical compliant system of an optical sensor unit; a – unloaded state, 1 – frame, 2 – moving part; b – loaded state

The moment M_L is unknown and is found using the expression from (3.22) and the boundary condition, that the tangent at $x = L$ remains parallel to the x-axis:

$$u_y'(L) = 0 . \tag{3.59}$$

This results in the following for the moment M_L:

$$M_L = \frac{1}{2} F_L L . \tag{3.60}$$

If the position of the joint along the length L is found by using the equation (3.26) and taking (3.60) into consideration, this position lies an infinite distance from the fixed end of the rod. This is due to the parallelism of the two ends of the rod, meaning there is no possible position of the joint that would fulfil the conditions (3.17)-(3.19) with only one joint. Therefore, one half of the rod is considered (Fig. 3.13 c); this is possible due to the point symmetry of the deformed rod shape. In order to determine the loads acting at the free end of this new rod, the expression for its internal moment (Fig. 3.13 b):

$$M_z(x) = M_L - F_L(L - x) \tag{3.61}$$

is expanded with the value for the moment M_L from (3.60) and the coordinate $x = L/2$. This results in an internal moment with a value of zero. Thus, there is no moment acting on the free end of the new modeled rod from Fig. 3.13 c. Only a force acts upon it, whose value is irrelevant and which is denoted with F_1. This is the load case $n = 2$, the corresponding values for δ and c_t for this modeled rod of length $L/2$ can thus be taken from Table 3.1 (Fig. 3.13 c):

$$\delta = \frac{2}{3}, \qquad c_t = \frac{8}{3}\frac{EI_z}{L}. \tag{3.62}$$

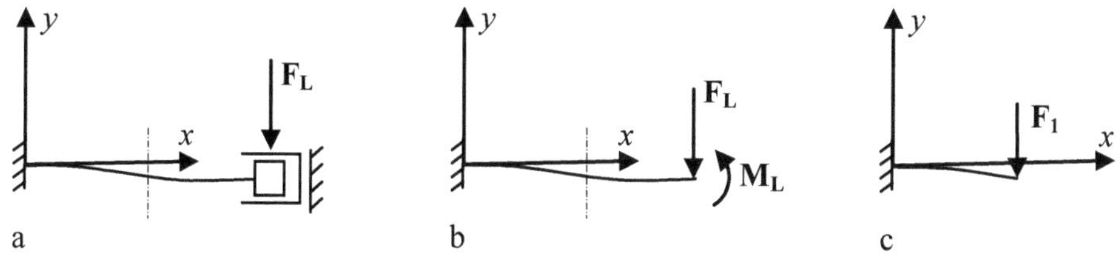

Fig. 3.13: Compliant element of the mechanical compliant system (Fig. 3.12) modeled as an elastic rod; a – schematic of the boundary conditions of the rod; b – depiction of the loads acting at the rod's end; c – half of the rod with length $L/2$ (used for reasons of symmetry)

Both halves are now combined into a single system (Fig. 3.14), resulting in a rigid-body model with two joints for the whole rod. The middle section of the system has a length $2L/3$, and the outer rigid-body parts connected with joints each has a length of $L/6$. The result of this rigid-body model is given in Fig. 3.14 c.

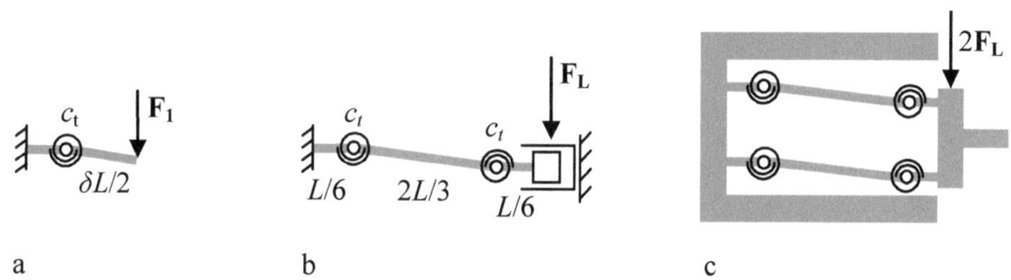

Fig. 3.14: Rigid-body model; a – for half of the compliant element; b – for the entire compliant element; c – rigid-body model of the entire mechanical compliant system

In this example, the position of the joints and the values of the spring constants (3.62) do not depend on the external loads. As long as the additional boundary condition in (3.59) is fulfilled, a special relationship (3.60) is formed between the acting force and the moment. Therefore, the rigid-body model in Fig. 3.14 c can be used for all parallel cascading compliant elements with a moving part that undergoes an approximately translational motion.

4 Modeling large deflections of curved rod-like structures

The subject of further experimentation is a thin elastic rod whose at least the deformed shape displays a curved axial line in three-dimensional space. An axial line is a line which connects the centroids of the cross-sections. In order to replicate the fluid-mechanical actuators, as well as the mechanisms or mechanism parts, in the modeling, hollow elastic rods will be considered alongside common curved rods. The hollow space of such rods is filled with a pressurized liquid. Moreover, elastic rods with embedded, length-constant yet limp elements will be including in the modeling. These elements will be displayed as a thread or as an inelastic strip, both bendable and unchanging in length. Should the pressure in the hollow space of the rods be increased or external loads be applied, the embedded elements maintain their constant length, while other parts of the system undergo stretching or compression. Under those assumptions, the deformations of fluid-mechanical actuators can be modeled.

The deflections of the rods are described on the basis of equilibrium conditions for a rod element (see [74] and [126]), yet the theory will be further developed so that the neutral axis given by the construction, an embedded length-constant and limp element, is also taken into consideration with the internal pressure and accordingly including in the equations (cf. [114]).

4.1 Assumptions for modeling

In mathematical models, a distinction is made between linear and non-linear models. Linear models deal either with a geometric linearity, in which only small displacements are permissible, or a material linearity, in which Hooke's law applies. Non-linear equations arise when large deflections in the rods are allowed, or when their material properties exhibit a non-linear character. Furthermore, curved rods, which undergo large deflections (geometrical non-linearity) yet exhibit linear material properties, will be modeled in three-dimensional space.

For the modeling of non-linear behavior, it is essential to distinguish the external loads. The direction of a *direction-constant load*, force or moment in large deflections stays consistent in the *Cartesian coordinate system* during a deformation. By contrast, a *follower load* follows the rod and remains its direction unchanged in a coordinate system fixed to the rod. Fig. 4.1 shows the difference in the deformation status when a follower force or a direction-constant force is applied, where each effective force has the same value. There are also other loads that cannot be assigned to either group.

The model equations result from the equilibrium of the rod elements. If the rod has a tube structure, the effect of the inner pressure, as well as the external forces and moments, is considered as external loads.

https://doi.org/10.1515/9783110479744-059

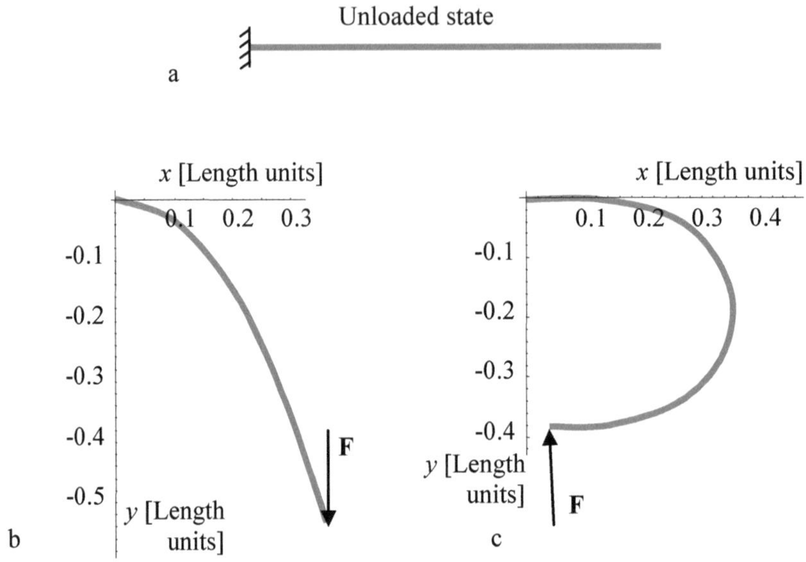

Fig. 4.1: Depiction of deformations under a direction-constant and follower force; a – rod with no load applied; b – deformation of a rod under load by a direction-constant force; c – deformation of a rod under load by a follower force; The forces in cases b and c have the same value

The following assumptions were made:

- a static problem is considered;
- thin rods are used: the cross-sectional dimensions are far smaller (ten times or more) then the length of the rods and their curvature radii, both in deformed and original states;
- the material of the rods fulfils *HOOKE's law*;
- *BERNOULLI's hypothesis* (also called *NAVIER's hypothesis*) applies: the cross-sections orthogonal to the axial line remain planar as load is applied;
- *SAINT-VENANT's principle* applies: the external force is applied evenly about the cross-section, as though the force was applied on the entire cross-section;
- the rod can be hollow and include an embedded, length-constant, bendable thread or strip.

Along with the assumption that the material of the rod fulfils HOOKE's law, solutions are only acceptable, as long as the maximal stress of the proportional limit (elastic limit) for the given material is not exceeded.

The *neutral axis* of the rod is either an axial line of the rod, or the actual fiber that is embedded into the wall of the rod as a length-constant yet bendable (bending stiffness are considered negligible – limp) thread or strip. An embedded thread is only considered in conjunction with a hollow space within the rod. Conversely, a hollow space without an embedded thread can be considered in the model. The origins of two vectors, $\mathbf{h}$ and $\mathbf{h}_T$, on the cross-section of the rod lie on the neutral axis; the first points towards the centroid of the cross-section of the hollow space and the second towards the centroid of the cross-section of the rod (Fig. 4.3). If neither a neutral axis (an embedded thread or strip) nor a hollow space is present in the body, the rod axis and the neutral axis coincide and the following applies: $\mathbf{h}_T = \mathbf{h} = \mathbf{0}$. The coordinate of a cross-section of the rod is described as parameter s. This is measured from the origin point of the Cartesian coordinate system along the neutral axis. The displacement from an unloaded to a loaded state of a rod is given as parameter $\mathbf{u}$ (Fig. 4.2).

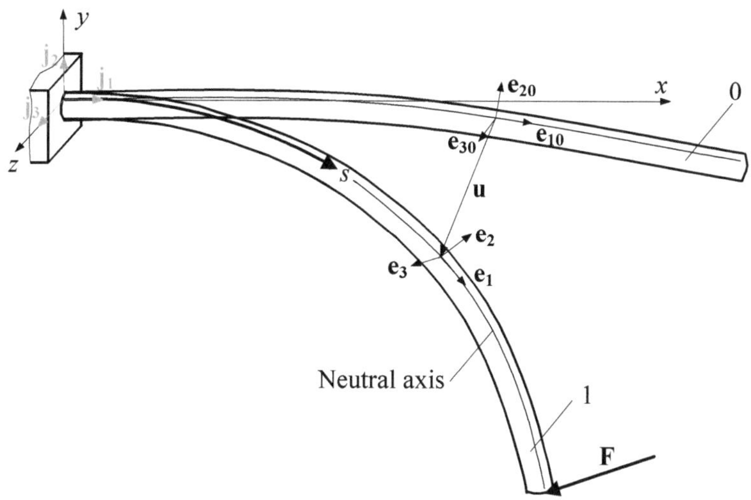

Fig. 4.2: Deformation of a rod: 0 – original, unloaded state, 1 – deformed, loaded state

The left-hand end of the rod is fixed and a Cartesian coordinate system can be applied to it in such a way, that the x-axis coincides with the tangent of the neutral axis at the left-hand end. The three axes *x, y* and *z* form a right-handed coordinate system with the corresponding basis of *unit vectors* $\mathbf{j}_1$, $\mathbf{j}_2$ and $\mathbf{j}_3$. A unit vector has a value of 1 and gives the direction of the coordinate axis. A further coordinate system with unit vectors $\mathbf{e}_1$, $\mathbf{e}_2$ and $\mathbf{e}_3$ is fixed to the neutral axis of the loaded rod and is named the *attached coordinate system*. The vector $\mathbf{e}_1$ lies on a tangent to the neutral axis of the loaded rod (Fig. 4.2). The unit vector $\mathbf{e}_2$ runs parallel to one of the principal axes of the cross-section, respectively named $\mathbf{e}_{T2}$. If possible, $\mathbf{e}_2$ should pass through the centroid of the cross-section of the hollow space, and be oriented opposite to the vector $\mathbf{h}$. The last unit vector $\mathbf{e}_3$ is oriented such that the *trihedron* $\{\mathbf{e}_1, \mathbf{e}_2, \mathbf{e}_3\}$ forms a right-handed coordinate system (Fig. 4.3). The unit vectors $\mathbf{e}_{T2}$ and $\mathbf{e}_{T3}$ are the principal axes of the rod cross-section.

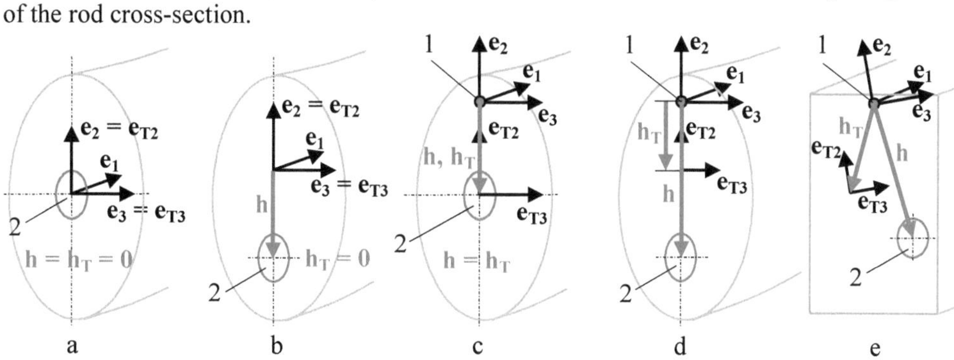

Fig. 4.3: Selected states of the hollow space and the thread on a cross-section; a – the centroid of the hollow space and of the rod coincide, without an embedded thread; b – the centroid of the hollow space and of the rod do not coincide, without an embedded thread; c – the centroid of the hollow space and of the rod coincide, with embedded thread; d – the straight line through the centroid of the hollow space and that of the rod intersects the embedded thread, **h** and $\mathbf{h}_T$ have the same direction; e – the straight line through the centroid of the hollow space and that of the rod does not intersect the embedded rod, **h** and $\mathbf{h}_T$ have differing directions; 1 – embedded thread, 2 – hollow space

The unit vectors $\mathbf{e}_{10}$, $\mathbf{e}_{20}$ and $\mathbf{e}_{30}$ are attached to the unloaded state of the rod and are introduced similarly to the trihedron $\{\mathbf{e}_1, \mathbf{e}_2, \mathbf{e}_3\}$. For general cases, the vectors $\mathbf{h}$ and $\mathbf{h}_T$ are represented as follows in an attached coordinate system:

$$\mathbf{h} = h_2\mathbf{e}_2 + h_3\mathbf{e}_3, \quad \mathbf{h}_T = h_{T2}\mathbf{e}_2 + h_{T3}\mathbf{e}_3. \tag{4.1}$$

4.2 Equilibrium conditions for a rod element

Deflections of a loaded rod are described in the attached coordinate system using equilibrium conditions for a rod element and for the fluid element, which are pressurized. The result are differential equations for forces and moments.

4.2.1 Equilibrium conditions in vector form

A rod element with the length ds is considered (Fig. 4.4). An internal force $-\mathbf{Q}$ acts at the cross-section of the rod element at the place with the coordinate s, which is a left-hand section. The force has three components: the longitudinal force Q_1 and the shear forces Q_2 und Q_3. A moment $\mathbf{M}$ also acts on this cross-section, also with three components, although M_1 is a torque and M_2 and M_3 are bending moments.

$$\mathbf{Q} = Q_1\mathbf{e}_1 + Q_2\mathbf{e}_2 + Q_3\mathbf{e}_3,$$
$$\mathbf{M} = M_1\mathbf{e}_1 + M_2\mathbf{e}_2 + M_3\mathbf{e}_3 \tag{4.2}$$

A force $\mathbf{Q} + d\mathbf{Q}$ and a moment $\mathbf{M} + d\mathbf{M}$ act at the other end of the rod element, at the right-hand section with the coordinate $s + ds$. The following illustration of these loads results from an expansion into a Taylor series about the parameter s:

$$\mathbf{Q}(s+ds) = \mathbf{Q}(s) + \frac{d\mathbf{Q}(s)}{ds}ds + \frac{d^2\mathbf{Q}(s)}{ds^2}ds^2 + \ldots. \tag{4.3}$$

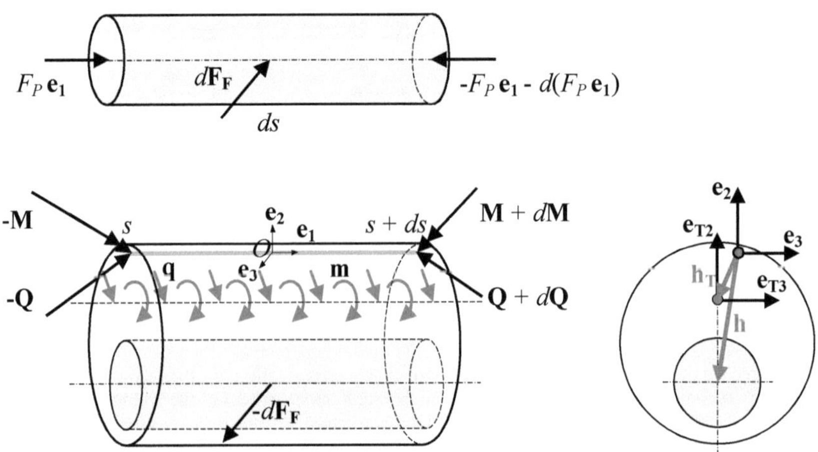

Fig. 4.4: A rod element filled with fluid; above: fluid element, below: rod element

Only the linear part of this expression is taken into account, as the other summands can be neglected. Thus, the following applies for the internal load at the point $s + ds$:

$$\mathbf{Q}(s+ds) = \mathbf{Q}(s) + d\mathbf{Q}(s)\,. \tag{4.4}$$

The expression for an internal moment can be found similarly. The hollow rod element (Fig. 4.4 below) and the corresponding fluid element (Fig. 4.4 above) are considered separately. The trihedron $\{\mathbf{e}_1, \mathbf{e}_2, \mathbf{e}_3\}$ is attached to the center point of the segment ds of the neutral axis. Similarly, the point of application of the forces and moments on a cross-section are considered together with the position of the neutral axis. A distributed force $\mathbf{q}$ also acts on the rod element, which can be taken as an example of gravitational force, as well as a distributed moment $\mathbf{m}$**,** both will be considered acting on the rod axis. The rod element and its filling are considered separately, equilibrium conditions for both systems are thus formulated. The equilibrium condition for the fluid element is:

$$-d(F_P\mathbf{e}_1) + d\mathbf{F}_\mathrm{F} = \mathbf{0}\,. \tag{4.5}$$

The force $\mathbf{F}_\mathrm{P}$, caused by the internal pressure p of the fluid, acts on the cross-sectional area A_P and remains orthogonal to it:

$$\mathbf{F}_P = F_P\mathbf{e}_1 = pA_P\mathbf{e}_1\,. \tag{4.6}$$

As p the difference between the pressure in the rod and the atmospheric pressure is denoted in the following. Due to the assumption that the cross-sectional dimensions of the rod are far smaller than its curvature radius, the pressure on the inner rod surface is balanced. The force $d\mathbf{F}_\mathrm{F}$ is a reactive force between the rod element and the fluid element. The equilibrium condition for the rod element is formulated as follows:

$$d\mathbf{Q} + \mathbf{q}ds - d\mathbf{F}_\mathrm{F} = \mathbf{0}\,. \tag{4.7}$$

The application of the force $d\mathbf{F}_\mathrm{F}$ from equation (4.5) in equation (4.7) and subsequent division of the result by ds results in a differential equation for the equilibrium of the force that act on the rod element:

$$\frac{d\mathbf{Q}}{ds} - \frac{d(F_P\mathbf{e}_1)}{ds} + \mathbf{q} = \mathbf{0}\,. \tag{4.8}$$

The moment equilibrium condition for the rod element is found with respect to the coordinate origin described in Fig. 4.4 below with the point O:

$$d\mathbf{M} + (ds\mathbf{e}_1 \times \mathbf{Q}) + (\mathbf{h}_\mathrm{T} \times \mathbf{q}ds) + \mathbf{m}ds - (\mathbf{h} \times d\mathbf{F}_\mathrm{P}) = \mathbf{0}\,. \tag{4.9}$$

In general cases, the vectors $\mathbf{h}$ and $\mathbf{h}_\mathrm{T}$ should be split into the components in the directions of $\mathbf{e}_2$ und $\mathbf{e}_3$. The parameter variations of higher orders, such as the product between $ds\mathbf{e}_1$ und $d\mathbf{Q}$ are neglected. The division of equation (4.9) by ds results in a differential equation for moments:

$$\frac{d\mathbf{M}}{ds} + (\mathbf{e}_1 \times \mathbf{Q}) + (\mathbf{h}_\mathrm{T} \times \mathbf{q}) + \mathbf{m} - (\mathbf{h} \times \frac{d\mathbf{F}_\mathrm{P}}{ds}) = \mathbf{0}\,. \tag{4.10}$$

If the unit vector $\mathbf{e}_2$ and the vectors $\mathbf{h}$ und $\mathbf{h}_\mathrm{T}$ lie on the same straight line (Fig. 4.3 a-d) and thus each has only one, usually negative, component, and the following applies:

$$\mathbf{h} = -h\mathbf{e}_2,\quad \mathbf{h}_\mathrm{T} = -h_T\mathbf{e}_2,\quad h_T \geq 0, h \geq 0\,, \tag{4.11}$$

then equation (4.10) is given as follows:

$$\frac{d\mathbf{M}}{ds}+(\mathbf{e_1}\times\mathbf{Q})-h_T(\mathbf{e_2}\times\mathbf{q})+\mathbf{m}+h(\mathbf{e_2}\times\frac{d\mathbf{F_P}}{ds})=\mathbf{0}\,. \tag{4.12}$$

The derivative of the force $\mathbf{F_P}$ in the last equation results from the derivative of the value of the cross-section area and the derivative of the unit vector $\mathbf{e_1}$, as long as the force p is distributed uniformly along the length of the rod (p = constant):

$$\frac{d\mathbf{F_P}}{ds}=\frac{d\left(F_P\mathbf{e_1}\right)}{ds}=pA_P'\mathbf{e_1}+F_P\frac{d\mathbf{e_1}}{ds},\quad A_P'=\frac{dA_P}{ds}\,. \tag{4.13}$$

Both equilibrium equations (4.8) and (4.10) are vector differential equations in the attached coordinate system. Accordingly, their basic vectors vary in direction during change of location s. Therefore, the derivatives of the forces and moments can be found as follows (compare with equation (4.2)):

$$\begin{aligned}\frac{d\mathbf{Q}}{ds}&=\frac{dQ_1}{ds}\mathbf{e_1}+\frac{d\mathbf{e_1}}{ds}Q_1+\frac{dQ_2}{ds}\mathbf{e_2}+\frac{d\mathbf{e_2}}{ds}Q_2+\frac{dQ_3}{ds}\mathbf{e_3}+\frac{d\mathbf{e_3}}{ds}Q_3,\\ \frac{d\mathbf{M}}{ds}&=\frac{dM_1}{ds}\mathbf{e_1}+\frac{d\mathbf{e_1}}{ds}M_1+\frac{dM_2}{ds}\mathbf{e_2}+\frac{d\mathbf{e_2}}{ds}M_2+\frac{dM_3}{ds}\mathbf{e_3}+\frac{d\mathbf{e_3}}{ds}M_3.\end{aligned} \tag{4.14}$$

Now, the derivatives of the unit vectors are to determinate, in order to conclude the derivation of the differential equations (4.8) and (4.10). This is the focus of the following section.

4.2.2 Derivative of the unit vectors

The results of the derivatives of the unit vectors $\mathbf{e_1}$, $\mathbf{e_2}$ and $\mathbf{e_3}$ with respect to the parameter s are also vectors, and can be split into components based on the directions $\mathbf{e_1}$, $\mathbf{e_2}$ and $\mathbf{e_3}$. These components are unknown and will be represented initially as κ_{ij} (i = 1,2,3; j = 1,2,3):

$$\begin{aligned}\frac{d\mathbf{e_1}}{ds}&=\kappa_{11}\mathbf{e_1}+\kappa_{12}\mathbf{e_2}+\kappa_{13}\mathbf{e_3},\\ \frac{d\mathbf{e_2}}{ds}&=\kappa_{21}\mathbf{e_1}+\kappa_{22}\mathbf{e_2}+\kappa_{23}\mathbf{e_3},\\ \frac{d\mathbf{e_3}}{ds}&=\kappa_{31}\mathbf{e_1}+\kappa_{32}\mathbf{e_2}+\kappa_{33}\mathbf{e_3}.\end{aligned} \tag{4.15}$$

The components κ_{ij} form a matrix of dimensions 3x3:

$$\underline{\boldsymbol{\kappa}}=\begin{pmatrix}\kappa_{11} & \kappa_{12} & \kappa_{13}\\ \kappa_{21} & \kappa_{22} & \kappa_{23}\\ \kappa_{31} & \kappa_{32} & \kappa_{33}\end{pmatrix}. \tag{4.16}$$

Subsequently, the properties of the components κ_{ij} will be identified, in order to determine the derivative of the unit vectors. A first property can be easily identified with equation (4.15), as the individual components can be found from the unit vectors and their derivatives:

$$\kappa_{ij} = \frac{d\mathbf{e}_i}{ds}\mathbf{e}_j\,. \tag{4.17}$$

Because the unit vectors are orthogonal to one another, they have the following property:

$$\mathbf{e}_i\mathbf{e}_j = 0, \quad i \neq j\,. \tag{4.18}$$

The derivative of equation (4.18) results in the following equation:

$$\frac{d\mathbf{e}_i}{ds}\mathbf{e}_j + \frac{d\mathbf{e}_j}{ds}\mathbf{e}_i = 0, \quad i \neq j\,. \tag{4.19}$$

Considering equation (4.17), the last equation can be written using component κ_{ij} as follows:

$$\kappa_{ij} = -\kappa_{ji}, \quad i \neq j\,. \tag{4.20}$$

Thus, of the nine unknown components, six constitute identical pairs concerning its value, reducing the unknowns by three. The first equation of (4.21) can be differentiated with respect to s in order to find the remaining properties of component κ_{ij}:

$$\begin{aligned} &\mathbf{e}_i\mathbf{e}_i = 1, \\ &\frac{d\mathbf{e}_i}{ds}\mathbf{e}_i + \frac{d\mathbf{e}_i}{ds}\mathbf{e}_i = 0. \end{aligned} \tag{4.21}$$

Subsequently, considering equation (4.17), three components of the matrix (4.16) can be found:

$$\kappa_{ii} = 0, \quad i = 1,2,3\,. \tag{4.22}$$

The matrix (4.16) takes on the following form, according to the discovered components shown in (4.20) and (4.22):

$$\underline{\boldsymbol{\kappa}} = \begin{pmatrix} 0 & \kappa_{12} & -\kappa_{31} \\ -\kappa_{12} & 0 & \kappa_{23} \\ \kappa_{31} & -\kappa_{23} & 0 \end{pmatrix} = \begin{pmatrix} 0 & \kappa_3 & -\kappa_2 \\ -\kappa_3 & 0 & \kappa_1 \\ \kappa_2 & -\kappa_1 & 0 \end{pmatrix}. \tag{4.23}$$

This matrix is thus skew-symmetric and is defined according to only three components. To simplify subsequent implementations, new designations will be introduced for the remaining components: $\kappa_1 = \kappa_{23}$, $\kappa_2 = \kappa_{31}$ and $\kappa_3 = \kappa_{12}$. Thus, the derivatives of the unit vectors are defined through only three components, κ_1, κ_2 and κ_3:

$$\begin{aligned} \frac{d\mathbf{e}_1}{ds} &= \kappa_3\mathbf{e}_2 - \kappa_2\mathbf{e}_3, \\ \frac{d\mathbf{e}_2}{ds} &= -\kappa_3\mathbf{e}_1 + \kappa_1\mathbf{e}_3, \\ \frac{d\mathbf{e}_3}{ds} &= \kappa_2\mathbf{e}_1 - \kappa_1\mathbf{e}_2. \end{aligned} \tag{4.24}$$

This system of equations can also be written in vector form. For such a derivative, a unit vector applies:

$$\frac{d\mathbf{e}_i}{ds} = \underline{\boldsymbol{\kappa}}^{T}\mathbf{e}_i . \tag{4.25}$$

Instead of a matrix, a vector can now be built from the three remaining components:

$$\boldsymbol{\kappa} = \kappa_1\mathbf{e}_1 + \kappa_2\mathbf{e}_2 + \kappa_3\mathbf{e}_3 . \tag{4.26}$$

Using this vector, the derivatives of the unit vectors can be written as follows:

$$\frac{d\mathbf{e}_i}{ds} = \boldsymbol{\kappa} \times \mathbf{e}_i . \tag{4.27}$$

This expression can be verified and thus proved for the three unit vectors $\mathbf{e}_1$, $\mathbf{e}_2$ and $\mathbf{e}_3$; a proof of equation (4.27) for $\mathbf{e}_1$ follows:

$$\frac{d\mathbf{e}_1}{ds} = \boldsymbol{\kappa} \times \mathbf{e}_1 = \left(\kappa_1\mathbf{e}_1 + \kappa_2\mathbf{e}_2 + \kappa_3\mathbf{e}_3\right) \times \mathbf{e}_1 = -\kappa_2\mathbf{e}_3 + \kappa_3\mathbf{e}_2 . \tag{4.28}$$

This expression for the derivative of $\mathbf{e}_1$ is confirmed by comparing it with the first equation from (4.24).

4.2.3 Natural coordinate system

Furthermore, the unit vectors $\mathbf{e}_{n1}$, $\mathbf{e}_{n2}$ and $\mathbf{e}_{n3}$ of a *natural coordinate system* for a spatial curve will be considered. Firstly, the vector $\mathbf{e}_{n1}$ lies on a tangent to this curve. The unit vector $\mathbf{e}_{n2}$ points to the center of its curvature, and $\mathbf{e}_{n3}$ is oriented so that the trihedron $\{\mathbf{e}_{n1}, \mathbf{e}_{n2}, \mathbf{e}_{n3}\}$ forms a right-handed coordinate system. This is a special case of the attached coordinate system, which was introduced in Section 4.1. Thus, the equation system (4.24) also applies for $\mathbf{e}_{ni}$ ($i = 1,2,3$), although the components for the natural coordinate system will be labelled κ_{ni}, in order to distinguish them from the components κ_i:

$$\begin{aligned}
\frac{d\mathbf{e}_{n1}}{ds} &= \kappa_{n3}\mathbf{e}_{n2} - \kappa_{n2}\mathbf{e}_{n3},\\
\frac{d\mathbf{e}_{n2}}{ds} &= -\kappa_{n3}\mathbf{e}_{n1} + \kappa_{n1}\mathbf{e}_{n3},\\
\frac{d\mathbf{e}_{n3}}{ds} &= \kappa_{n2}\mathbf{e}_{n1} - \kappa_{n1}\mathbf{e}_{n2}.
\end{aligned} \tag{4.29}$$

The trihedron of the natural coordinate system is used to explain the geometric meaning of the components κ_{ni}, in order to obtain the information for the components κ_i. In the later considerations, new designations and parameters will be required. The surfaces each formed from two unit vectors will be labelled γ_{n1}, γ_{n2} and γ_{n3}, where the first surface lies orthogonal to $\mathbf{e}_{n1}$, the second to $\mathbf{e}_{n2}$ and the last to $\mathbf{e}_{n3}$. Additionally, three angles for the rotation about the unit vectors will be introduced and labelled θ_{ni}, $i = 1,2,3$ (Fig. 4.5 a).

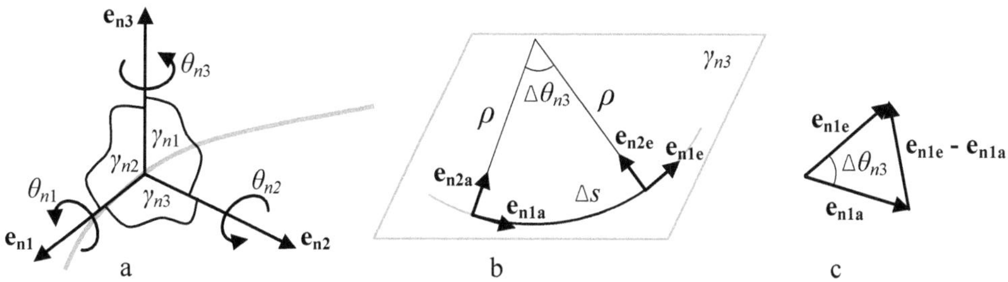

Fig. 4.5: Natural coordinate system with labels; a – trihedron, plains and angles; b – movement of the trihedron from the start to end of the curve element Δs with a curvature radius ρ; c – difference of the unit vectors (see equation (4.30))

Subsequently, the plane γ_{n3} will be considered, containing of a curve element with a length of Δs. When the trihedron is displaced from the start to the end of the curve element, the trihedron undergoes a rotation by the angle $\Delta\theta_{n3}$. This situation is given in Fig. 4.5 b. The unit vector $\mathbf{e}_{n1}$ also changes its direction. Its derivative also differs from zero and can be written according to its mathematical definition considering the geometrical relationships:

$$\frac{d\mathbf{e}_{n1}}{ds} = \lim_{\Delta s \to 0} \frac{\Delta \mathbf{e}_{n1}}{\Delta s} = \lim_{\Delta s \to 0} \frac{\mathbf{e}_{n1e} - \mathbf{e}_{n1a}}{\Delta s} = \lim_{\Delta s \to 0} \frac{\Delta\theta_{n3}\mathbf{e}_{n2}}{\Delta s}. \tag{4.30}$$

The unit vector $\mathbf{e}_{n1}$ is indicated through the parameters $\mathbf{e}_{n1a}$ and $\mathbf{e}_{n1e}$ (Fig. 4.5 c), referring to the start and end of the length Δs. A vector that forms the difference $\mathbf{e}_{n1e} - \mathbf{e}_{n1a}$ is directed towards the center of curvature and therefore has a direction $\mathbf{e}_{n2}$. According to the following geometric context:

$$\frac{\Delta\theta_{n3}}{\Delta s} = \frac{1}{\rho}, \tag{4.31}$$

the derivative of the unit vector is written as follows:

$$\frac{d\mathbf{e}_{n1}}{ds} = \lim_{\Delta s \to 0} \frac{1}{\rho}\mathbf{e}_{n2} = \frac{1}{\rho}\mathbf{e}_{n2}. \tag{4.32}$$

When this expression is compared with the first equation from (4.29), the values and thus the meanings of two components of this equation can be found:

$$\begin{aligned} \kappa_{n3} &= \frac{1}{\rho}, \\ \kappa_{n2} &= 0. \end{aligned} \tag{4.33}$$

The component κ_{n3} is dependent on the radius of the curve and is named the *curvature*. The component κ_{n2} has the value of zero in the natural coordinate system. To answer the question of the geometric meaning of the component κ_{n1}, the last equation of equation system (4.29) is considered. For this purpose, the displacement of the trihedron $\{\mathbf{e}_{n1}, \mathbf{e}_{n2}, \mathbf{e}_{n3}\}$ along the curve element Δs is observed, although its rotation on the plane γ_{n1} is the focus here (Fig. 4.6 a). The derivative of the unit vector $\mathbf{e}_{n3}$, supported by the graphics from Fig. 4.6 b-c, is written as follows:

$$\frac{d\mathbf{e}_{\mathbf{n3}}}{ds} = \lim_{\Delta s \to 0} \frac{\Delta \mathbf{e}_{\mathbf{n3}}}{\Delta s} = \lim_{\Delta s \to 0} \frac{\mathbf{e}_{\mathbf{n3e}} - \mathbf{e}_{\mathbf{n3a}}}{\Delta s} = \lim_{\Delta s \to 0} \left(-\frac{\Delta \theta_{n1} \mathbf{e}_{\mathbf{n2}}}{\Delta s} \right) = -\frac{d\theta_{n1}}{ds} \mathbf{e}_{\mathbf{n2}} . \tag{4.34}$$

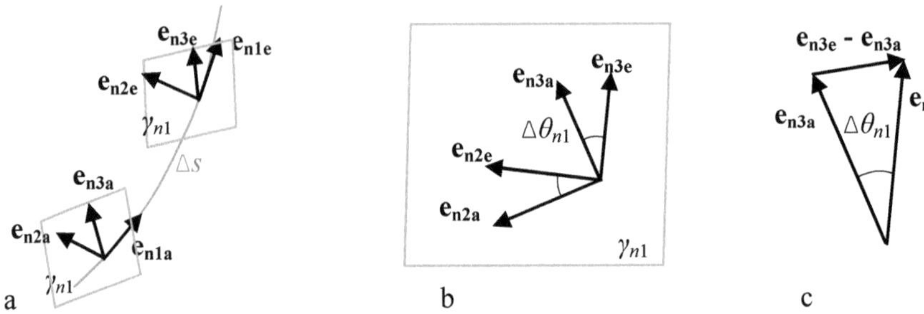

Fig. 4.6: Referring to the unit vector $\mathbf{e}_{\mathbf{n3}}$; a – displacement of trihedron along the curve; b – rotation of the unit vector $\mathbf{e}_{\mathbf{n3}}$ on the plane γ_{n1}; c – difference of the unit vectors (see equation (4.34))

A vector formed from the difference between $\mathbf{e}_{\mathbf{n3e}}$ and $\mathbf{e}_{\mathbf{n3a}}$ points in the opposite direction to $\mathbf{e}_{\mathbf{n2}}$, hence the minus sign in the previous expression. Using the equation (4.34) and considering the last equation from the equation system (4.29), the meaning of the component κ_{n1} is given. This reflects the rotation of the trihedron about the tangent, or about the unit vector $\mathbf{e}_{\mathbf{n1}}$; hereafter named the *twist* or second curvature of a spatial curve.

$$\kappa_{n1} = \frac{d\theta_{n1}}{ds} \tag{4.35}$$

The vector $\boldsymbol{\kappa}_{\mathbf{n}}$, formed from the corresponding components:

$$\boldsymbol{\kappa}_{\mathbf{n}} = \kappa_{n1}\mathbf{e}_{\mathbf{n1}} + 0\mathbf{e}_{\mathbf{n2}} + \kappa_{n3}\mathbf{e}_{\mathbf{n3}}$$

is named the *DARBOUX vector* (see also [1]). Using the discovered components: curvature and twist according to (4.33) and (4.35), the equation system can be newly written as (4.29):

$$\begin{aligned}
\frac{d\mathbf{e}_{\mathbf{n1}}}{ds} &= \kappa_{n3}\mathbf{e}_{\mathbf{n2}}, \\
\frac{d\mathbf{e}_{\mathbf{n2}}}{ds} &= -\kappa_{n3}\mathbf{e}_{\mathbf{n1}} + \kappa_{n1}\mathbf{e}_{\mathbf{n3}}, \\
\frac{d\mathbf{e}_{\mathbf{n3}}}{ds} &= -\kappa_{n1}\mathbf{e}_{\mathbf{n2}}.
\end{aligned} \tag{4.36}$$

These equations are named *FRENET formulas* and the trihedron $\{\mathbf{e}_{\mathbf{n1}}, \mathbf{e}_{\mathbf{n2}}, \mathbf{e}_{\mathbf{n3}}\}$ is called FRENET trihedron. In kinematics, equations of this form (4.36) are used to describe the movement of a point, or a center of gravity of a rigid body along a curve, as described in [1].

4.2.4 Relationship between the natural coordinate system and attached coordinate system

When the unit vectors of the natural coordinate systems, with the unit vectors $\mathbf{e}_{\mathbf{n1}}$, $\mathbf{e}_{\mathbf{n2}}$ and $\mathbf{e}_{\mathbf{n3}}$, and the attached coordinate systems, with the unit vectors $\mathbf{e}_1$, $\mathbf{e}_2$ and $\mathbf{e}_3$, are compared, the following can be concluded. The vectors $\mathbf{e}_{\mathbf{n1}}$ and $\mathbf{e}_1$ always correspond relating to direction, because these lie on the tangent of the curve or the neutral axis. Though, these are twisted versus each other by an angle $\theta_{(1)}$. The vector pairs $\mathbf{e}_{n2}$ and $\mathbf{e}_{n3}$, as well as $\mathbf{e}_2$ and $\mathbf{e}_3$ lie on a plane γ_1, because both pares are orthogonal to $\mathbf{e}_{n1}$ or $\mathbf{e}_1$. These vector pairs, $\mathbf{e}_{n2}$, $\mathbf{e}_{n3}$ and $\mathbf{e}_2$, $\mathbf{e}_3$, can exhibit different directions (Fig. 4.7 a), which differ about the angle $\theta_{(1)}$. The vector $\boldsymbol{\kappa}_n$ is defined in the natural coordinate system and can be written as follows:

$$\boldsymbol{\kappa}_{\mathbf{n}} = \kappa_{n1}\mathbf{e}_{\mathbf{n1}} + \kappa_{n3}\mathbf{e}_{\mathbf{n3}}\,. \tag{4.37}$$

Similarly, both unit vectors in directions $\mathbf{e}_1$, $\mathbf{e}_2$ and $\mathbf{e}_3$ can be written as:

$$\begin{aligned} \mathbf{e}_{\mathbf{n1}} &= \mathbf{e}_{\mathbf{1}}, \\ \mathbf{e}_{\mathbf{n3}} &= \sin\theta_{(1)}\mathbf{e}_{\mathbf{2}} + \cos\theta_{(1)}\mathbf{e}_{\mathbf{3}}. \end{aligned} \tag{4.38}$$

The vector $\boldsymbol{\kappa}$ in the attached coordinate system corresponds to the vector $\boldsymbol{\kappa}_n$ in the natural coordinate system and can be written as follows:

$$\boldsymbol{\kappa} = \kappa_1\mathbf{e}_{\mathbf{1}} + \kappa_{n3}\sin\theta_{(1)}\mathbf{e}_{\mathbf{2}} + \kappa_{n3}\cos\theta_{(1)}\mathbf{e}_{\mathbf{3}}\,, \tag{4.39}$$

wherein its three components differ from zero:

$$\begin{aligned} \kappa_1 &= \frac{d(\theta_{n1} + \theta_{(1)})}{ds} = \kappa_{n1} + \frac{d\theta_{(1)}}{ds}, \\ \kappa_2 &= \kappa_{n3}\sin\theta_{(1)}, \\ \kappa_3 &= \kappa_{n3}\cos\theta_{(1)}. \end{aligned} \tag{4.40}$$

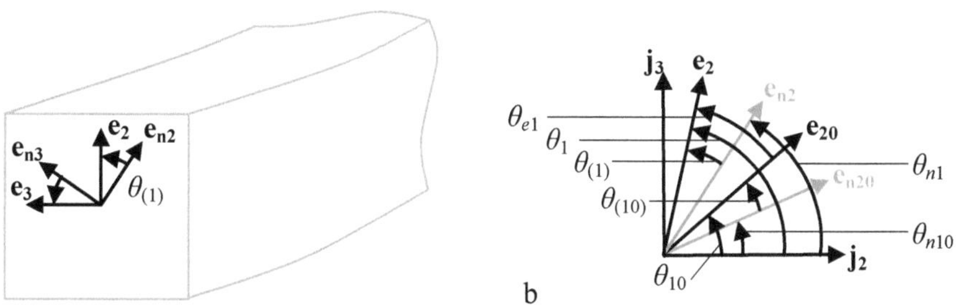

Fig. 4.7: Position and designations of the unit vectors; a – vector pairs $\mathbf{e}_{n2}$, $\mathbf{e}_{n3}$ and $\mathbf{e}_2$, $\mathbf{e}_3$ in the cross-sectional area of a rod; b – designations for the indices of the angles θ in the plane γ_1 for the following situations: $\mathbf{e}_{n20}$ – for natural coordinate system, unloaded state, $\mathbf{e}_{n2}$ – for natural coordinate system, loaded state, $\mathbf{e}_{20}$ – for attached coordinate system, unloaded state, $\mathbf{e}_2$ – for attached coordinate system, loaded state, $\mathbf{j}_2, \mathbf{j}_3$ – for Cartesian coordinate system

The first component κ_1 consists of the curve twist and the twisting between the attached coordinate system and the natural coordinate system. This fact can be carried out through the sum of the mentioned quantities. The second and third components, κ_2 and κ_3, are the corresponding projection of the curvature κ_{n3} on $\mathbf{e}_2$ and $\mathbf{e}_3$.

The derivative of the unit vectors of the attached coordinate system will later be found according to (4.27). The designations from Fig. 4.5 a are also used for the trihedrons $\{\mathbf{e}_1, \mathbf{e}_2, \mathbf{e}_3\}$ and $\{\mathbf{e}_{10}, \mathbf{e}_{20}, \mathbf{e}_{30}\}$, although the indices are adapted, i.e. index "0" is used for the original state of the rod. In Fig. 4.7 b, the designations for the indices of the angles θ in the plane γ_1 are shown. The labels for the angle "θ", as well as for the plane "γ" are also used and provided with the respective indices.

4.2.5 Further development of the equilibrium equations

The derivatives of the forces and moments for the equilibrium equations (4.8) and (4.10) can now be written according to (4.14) by using expression (4.27) for the derivatives of the unit vectors.

$$\begin{aligned}\frac{d\mathbf{Q}}{ds} &= \frac{dQ_1}{ds}\mathbf{e}_1 + \frac{dQ_2}{ds}\mathbf{e}_2 + \frac{dQ_3}{ds}\mathbf{e}_3 + \boldsymbol{\kappa}\times\mathbf{Q} = \frac{d'\mathbf{Q}}{ds} + \boldsymbol{\kappa}\times\mathbf{Q},\\ \frac{d\mathbf{M}}{ds} &= \frac{dM_1}{ds}\mathbf{e}_1 + \frac{dM_2}{ds}\mathbf{e}_2 + \frac{dM_3}{ds}\mathbf{e}_3 + \boldsymbol{\kappa}\times\mathbf{M} = \frac{d'\mathbf{M}}{ds} + \boldsymbol{\kappa}\times\mathbf{M}\end{aligned} \tag{4.41}$$

Here, the sum of the derivatives of the values of the forces or moments are defined as *local derivatives* and written with a “stroke”, to distinguish them from the real derivatives:

$$\begin{aligned}\frac{d'\mathbf{Q}}{ds} &= \frac{dQ_1}{ds}\mathbf{e}_1 + \frac{dQ_2}{ds}\mathbf{e}_2 + \frac{dQ_3}{ds}\mathbf{e}_3,\\ \frac{d'\mathbf{M}}{ds} &= \frac{dM_1}{ds}\mathbf{e}_1 + \frac{dM_2}{ds}\mathbf{e}_2 + \frac{dM_3}{ds}\mathbf{e}_3.\end{aligned} \tag{4.42}$$

Using this relationship and considering expression (4.13), the equilibrium equations for forces (4.8) can be written as follows:

$$\frac{d'\mathbf{Q}}{ds} + \boldsymbol{\kappa}\times(\mathbf{Q}-\mathbf{F}_\mathbf{P}) - pA_P'\mathbf{e}_1 + \mathbf{q} = \mathbf{0}\,. \tag{4.43}$$

To adapt the equilibrium equations for moments (4.10), the last term of these equations is initially written as follows:

$$(\mathbf{h}\times\frac{d\mathbf{F}_\mathbf{P}}{ds}) = \mathbf{h}\times(A_P' p\mathbf{e}_1 + F_P\frac{d\mathbf{e}_1}{ds}) = pA_P'(\mathbf{h}\times\mathbf{e}_1) + pA_P(\mathbf{h}\times(\boldsymbol{\kappa}\times\mathbf{e}_1))\,. \tag{4.44}$$

In cases where vector **h** only has one component in the direction $\mathbf{e}_2$ (see equations (4.11)), the last term of equation (4.12) is simplified as in (4.45).

$$(h\mathbf{e}_2\times\frac{d\mathbf{F}_\mathbf{P}}{ds}) = h\mathbf{e}_2\times(A_P' p\mathbf{e}_1 + F_P\frac{d\mathbf{e}_1}{ds}) = -hp(A_P\kappa_2\mathbf{e}_1 + A_P'\mathbf{e}_3) \tag{4.45}$$

The term according to (4.44) can now be applied to the moment equilibrium equation (4.10); the derivative of the moment according to expression (4.41) can also be applied:

$$\begin{aligned}&\frac{d'\mathbf{M}}{ds} + \boldsymbol{\kappa}\times\mathbf{M} + (\mathbf{e}_1\times\mathbf{Q}) + (\mathbf{h}_\mathrm{T}\times\mathbf{q})\\ &\quad - pA_P'(\mathbf{h}\times\mathbf{e}_1) - pA_P(\mathbf{h}\times(\boldsymbol{\kappa}\times\mathbf{e}_1)) + \mathbf{m} = \mathbf{0}.\end{aligned} \tag{4.46}$$

Equations (4.43) and (4.46) are differential equilibrium conditions for a deformed rod.

4.3 Inclusion of material properties

A stress $\boldsymbol{\sigma}$ occurs on the cross-sections of a loaded rod, which can be split into three components according to the trihedron $\{\mathbf{e}_1, \mathbf{e}_2, \mathbf{e}_3\}$. This stress consists of a normal stress σ_1, orthogonal to the cross-section and a shear stress $\boldsymbol{\tau}$ that lies on the cross-sectional plane and contains two components, τ_2 and τ_3. The indices illustrate the directions of the stress components (Fig. 4.8).

$$\boldsymbol{\sigma} = \sigma_1 \mathbf{e}_1 + \tau_2 \mathbf{e}_2 + \tau_3 \mathbf{e}_3 \tag{4.47}$$

A radius vector $\mathbf{r}$ describes the position of any point P on the cross-section of a deformed rod and has two components:

$$\mathbf{r} = r_2 \mathbf{e}_2 + r_3 \mathbf{e}_3 \,. \tag{4.48}$$

A cross-sectional moment $\mathbf{M}$ which also appears in equilibrium equation (4.46), is constituted by the stress $\boldsymbol{\sigma}$:

$$\mathbf{M} = \int (\mathbf{r} \times \boldsymbol{\sigma}) dA = \int \left((r_2 \mathbf{e}_2 + r_3 \mathbf{e}_3) \times (\sigma_1 \mathbf{e}_1 + \tau_2 \mathbf{e}_2 + \tau_3 \mathbf{e}_3) \right) dA \,. \tag{4.49}$$

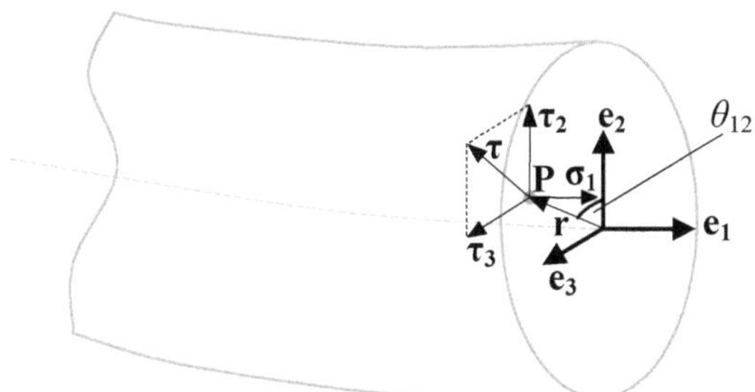

Fig. 4.8: Components of the stress $\boldsymbol{\sigma}$ on the cross-section; normal stress σ_1 and shear stresses τ_2 and τ_3 for a point determined by a radius vector $\mathbf{r}$ on the cross-section

The individual components of the moment can thus be written as follows:

$$\begin{aligned} M_1 &= \int (r_2 \tau_3 - r_3 \tau_2) dA, \\ M_2 &= \int (r_3 \sigma_1) dA, \\ M_3 &= \int (-r_2 \sigma_1) dA. \end{aligned} \tag{4.50}$$

Initially, the first moment M_1 is considered, in order to relate this to the deformation of the rod using HOOKE's law. This moment induces the torsion of the rod, as shown with the example of a rod element in Fig. 4.9. The corresponding shear stress depends linearly on the angle φ, which results by the proportionality constant G, a shear modulus, according to HOOKE's law as follows:

$$\tau = G \varphi \,. \tag{4.51}$$

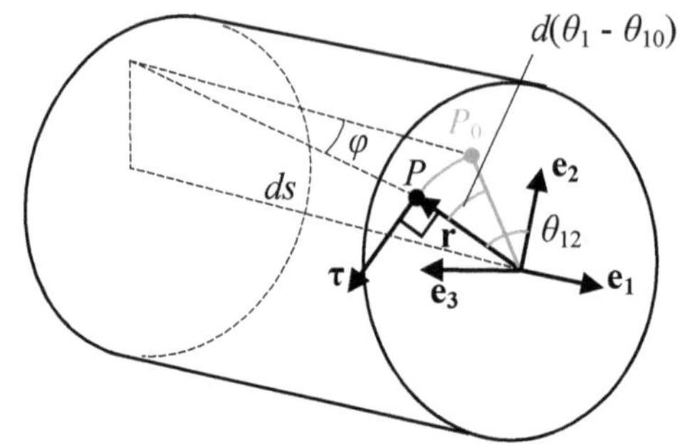

Fig. 4.9: Depiction of the shear stress during the torsion of the rod element in the direction $\mathbf{e}_1$

The angle $d(\theta_1 - \theta_{10})$ stands for the torsion of the rod element with a length ds under external load. The following can be concluded from the geometrical constellation shown in Fig. 4.9:

$$d(\theta_1 - \theta_{10})r = \varphi ds\,. \tag{4.52}$$

The angle φ from the last equation is applied to equation in (4.51):

$$\tau = Gr\frac{d(\theta_1 - \theta_{10})}{ds}\,. \tag{4.53}$$

The shear stress $\boldsymbol{\tau}$ contains two components relating both directions $\mathbf{e}_2$ and $\mathbf{e}_3$ on the cross-section:

$$\boldsymbol{\tau} = \tau_2\mathbf{e}_2 + \tau_3\mathbf{e}_3 = -\tau\sin\theta_{12}\mathbf{e}_2 + \tau\cos\theta_{12}\mathbf{e}_3\,. \tag{4.54}$$

The angle θ_{12} is an angle between the radius vector $\mathbf{r}$ to the chosen point P and the $\mathbf{e}_2$-direction on the cross-sectional plane of the rod (Fig. 4.8 and Fig. 4.9):

$$\cos\theta_{12} = \frac{r_2}{r},\qquad \sin\theta_{12} = \frac{r_3}{r}. \tag{4.55}$$

The first moment of equation system (4.50) can be written as follows, taking into consideration equations (4.53)-(4.55):

$$M_1 = \int G\frac{d(\theta_1 - \theta_{10})}{ds}(r_2^2 + r_3^2)dA\,. \tag{4.56}$$

The angle $(\theta_1 - \theta_{10})$ for the torsion of the rod can be given through four angles according to notations in Fig. 4.7 b:

$$\theta_1 - \theta_{10} = \theta_{n1} + \theta_{(1)} - (\theta_{n10} + \theta_{(10)})\,. \tag{4.57}$$

The derivative of this angle is noted in the following form, considering expression (4.40) for κ_1:

$$\frac{d(\theta_1 - \theta_{10})}{ds} = \frac{d\theta_{n1}}{ds} + \frac{d\theta_{(1)}}{ds} - \left(\frac{d\theta_{n10}}{ds} + \frac{d\theta_{(10)}}{ds}\right) = \kappa_1 - \kappa_{(10)}\,. \tag{4.58}$$

A new parameter, $\kappa_{(10)}$, is introduced here:

$$\kappa_{(10)} = \frac{d\theta_{n10}}{ds} + \frac{d\theta_{(10)}}{ds} = \kappa_{n10} + \frac{d\theta_{(10)}}{ds} = \frac{d\theta_{10}}{ds}, \quad (4.59)$$

which characterizes a torsion of the attached coordinate system for a still unloaded state of the rod. The parameter κ_{n10} is the twist of the neutral axis of the unloaded rod in the natural coordinate system. It is also considered that

$$I_1 = \int (r_2^2 + r_3^2) dA \quad (4.60)$$

represents a polar moment of area of the rod's cross-sectional area about $\mathbf{e}_1$. The torsional moment relates to the deformation of the rod according to (4.56) as follows:

$$M_1 = GI_1(\kappa_1 - \kappa_{(10)}). \quad (4.61)$$

The moment M_1 induces the change of the twist of the rod.

The other moment from equation system (4.50), the bending moment M_2, should also be found in relation to the deformation of the rod. Thus, HOOKE's law is applied, according to which the normal stress σ_1 is equal to the product of the stretching of the rod and a proportionality constant, the YOUNG's modulus E:

$$\sigma_1 = E\varepsilon. \quad (4.62)$$

The strain of the rod ε is considered using the change in length of an axis that is removed from the neutral axis about r_{n2}. Its length before deformation is given as ds_{f0}, and after a deformation as ds_f. The deformation of the rod is now considered on the plane γ_{n3}, on which both $\mathbf{e}_{n1}$ and $\mathbf{e}_{n2}$ lie. The stretching of the axis is composed as follows (Fig. 4.10 a).

$$\varepsilon = \frac{ds_f - ds_{f0}}{ds_{f0}} = \frac{(\rho - r_{n2})d\theta_{n3} - (\rho_0 - r_{n20})d\theta_{n30}}{(\rho_0 - r_{n20})d\theta_{n30}} \quad (4.63)$$

It should be noted that the length of the neutral axis remains constant, therefore the following expression applies:

$$\rho_0 d\theta_{n30} = \rho d\theta_{n3}. \quad (4.64)$$

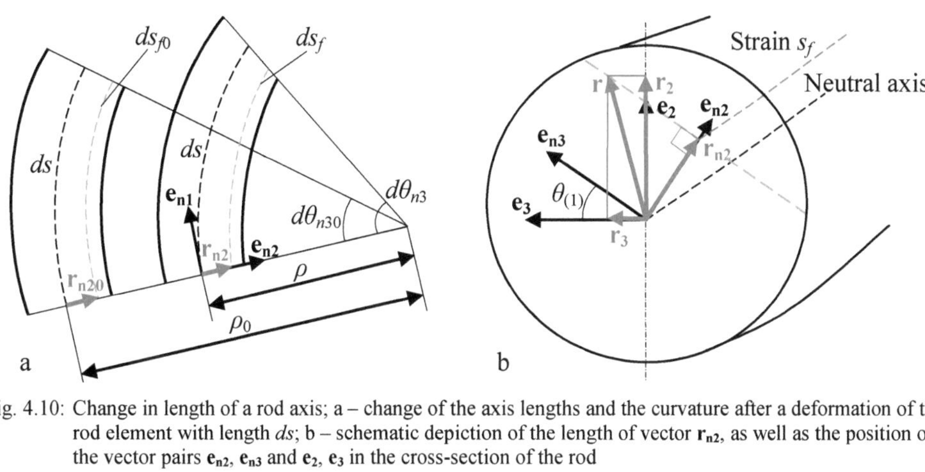

Fig. 4.10: Change in length of a rod axis; a – change of the axis lengths and the curvature after a deformation of the rod element with length ds; b – schematic depiction of the length of vector $\mathbf{r}_{n2}$, as well as the position of the vector pairs $\mathbf{e}_{n2}$, $\mathbf{e}_{n3}$ and $\mathbf{e}_2$, $\mathbf{e}_3$ in the cross-section of the rod

The strain (4.63) can now be written, taking into consideration the relationship (4.64) and the fact that r_{n20} is far smaller than ρ_0:

$$\varepsilon = \frac{1}{\rho_0 - r_{n20}} \left(r_{n20} - \frac{\rho_0}{\rho} r_{n2} \right) \approx r_{n20} \frac{1}{\rho_0} - r_{n2} \frac{1}{\rho}. \tag{4.65}$$

The values of the vectors $\mathbf{r_{n2}}$ and $\mathbf{r_{n20}}$ remain the same and can be replaced with the value of the components $\mathbf{r_2}$ and $\mathbf{r_3}$ of the radius vector $\mathbf{r}$ in the attached coordinate system of the deformed rod (Fig. 4.10 b):

$$r_{n2} = r_{n20} = r_2 \cos\theta_{(1)} - r_3 \sin\theta_{(1)}. \tag{4.66}$$

Considering (4.40), the following expressions for κ_2 and κ_3 can be used in further equations:

$$\kappa_2 = \frac{1}{\rho} \sin\theta_{(1)}, \quad \kappa_3 = \frac{1}{\rho} \cos\theta_{(1)}. \tag{4.67}$$

The parameters r_{n2} and r_{n20} are replaced according to equation (4.66) to expand the description of the strain:

$$\varepsilon = (r_2 \cos\theta_{(1)} - r_3 \sin\theta_{(1)}) \left(\frac{1}{\rho_0} - \frac{1}{\rho} \right). \tag{4.68}$$

New parameters, $\kappa_{(20)}$ and $\kappa_{(30)}$, for the following expressions is introduced:

$$\kappa_{(20)} = \frac{1}{\rho_0} \sin\theta_{(1)}, \quad \kappa_{(30)} = \frac{1}{\rho_0} \cos\theta_{(1)}. \tag{4.69}$$

Finally, the strain takes on the following form:

$$\varepsilon = r_3 \left(\kappa_2 - \kappa_{(20)} \right) - r_2 \left(\kappa_3 - \kappa_{(30)} \right). \tag{4.70}$$

The strain from equation (4.70) is applied to equation (4.62) for the stress that is then used for the bending moment M_2 from equation system (4.50):

$$M_2 = EI_2 (\kappa_2 - \kappa_{(20)}) + EI_{23} (\kappa_3 - \kappa_{(30)}). \tag{4.71}$$

The parameter I_2 is a second moment of area about an axis that coincides with the unit vector $\mathbf{e_2}$

$$I_2 = \int r_3^2 \, dA \tag{4.72}$$

and I_{23} is the product moment of area

$$I_{23} = -\int r_2 r_3 \, dA. \tag{4.73}$$

Upon substituting the equation (4.70) into equation (4.62) and the result then used for the bending moment M_3 from equation system (4.50), the latter moment takes on the following form:

$$M_3 = EI_3 (\kappa_3 - \kappa_{(30)}) + EI_{23} (\kappa_2 - \kappa_{(20)}). \tag{4.74}$$

Here, the second moment of area I_3 about an axis that coincides with $\mathbf{e}_3$ is expressed as follows:

$$I_3 = \int r_2^2 dA . \tag{4.75}$$

The difference $(\kappa_2 - \kappa_{(20)})$ represents the change in the curvature of the neutral axis of the rod, which is projected as a component of the $\mathbf{e}_{n3}$-direction (compare equations (4.37)-(4.40)) onto the direction $\mathbf{e}_2$.

$$\kappa_2 - \kappa_{(20)} = \left(\frac{1}{\rho} - \frac{1}{\rho_0}\right)\sin\theta_{(1)} \tag{4.76}$$

Similarly, the expression $(\kappa_3 - \kappa_{(30)})$ should be seen as a projection of the change in curvature of the neutral axis in the direction $\mathbf{e}_3$.

$$\kappa_3 - \kappa_{(30)} = \left(\frac{1}{\rho} - \frac{1}{\rho_0}\right)\cos\theta_{(1)} \tag{4.77}$$

Three equations result from this, which relate the external loads with the shape of the loaded rod, described with the components of the vector $\boldsymbol{\kappa}$:

$$\begin{aligned} M_1 &= GI_1(\kappa_1 - \kappa_{(10)}), \\ M_2 &= EI_2(\kappa_2 - \kappa_{(20)}) + EI_{23}(\kappa_3 - \kappa_{(30)}), \\ M_3 &= EI_3(\kappa_3 - \kappa_{(30)}) + EI_{23}(\kappa_2 - \kappa_{(20)}). \end{aligned} \tag{4.78}$$

If a stiffness matrix is introduced as follows:

$$\underline{\mathbf{S}} = \begin{pmatrix} GI_1 & 0 & 0 \\ 0 & EI_2 & EI_{23} \\ 0 & EI_{23} & EI_3 \end{pmatrix}, \tag{4.79}$$

then the equations from (4.78) can be written as an equation in vector form:

$$\mathbf{M} = \underline{\mathbf{S}}(\boldsymbol{\kappa} - \boldsymbol{\kappa}_{(0)}), \tag{4.80}$$

whereby the parameters $\mathbf{M}$, $\boldsymbol{\kappa}$ and $\boldsymbol{\kappa}_{(0)}$ are used as column vectors:

$$\mathbf{M} = \begin{pmatrix} M_1 \\ M_2 \\ M_3 \end{pmatrix}, \quad \boldsymbol{\kappa} = \begin{pmatrix} \kappa_1 \\ \kappa_2 \\ \kappa_3 \end{pmatrix}, \quad \boldsymbol{\kappa}_{(0)} = \begin{pmatrix} \kappa_{(10)} \\ \kappa_{(20)} \\ \kappa_{(30)} \end{pmatrix} = \begin{pmatrix} \dfrac{d\theta_{n10}}{ds} + \dfrac{d\theta_{(10)}}{ds} \\ \dfrac{1}{\rho_0}\sin\theta_{(1)} \\ \dfrac{1}{\rho_0}\cos\theta_{(1)} \end{pmatrix}. \tag{4.81}$$

The elements κ_{i0} and $\kappa_{(i0)}$ are elements of the same vector, but in different coordinate systems, in basis $\{\mathbf{e}_{10}, \mathbf{e}_{20}, \mathbf{e}_{30}\}$ and basis $\{\mathbf{e}_1, \mathbf{e}_2, \mathbf{e}_3\}$, respectively. If the axes, which coincide with $\mathbf{e}_2$ and $\mathbf{e}_3$, are principal axes or one of these is parallel to one of the principal axis (as in Fig. 4.3 a-d) then the product moment of area I_{23} equals null. The scalar equations from (4.78) or the vector equation (4.80) represent a relationship between the loads and the resultant loaded shape of the rod. The components of the parameter $\boldsymbol{\kappa}$ defined the shape of the neutral axis, precisely as it describes the curvature and twist of a curve in the natural coordinate system.

4.4 Transformation of unit vectors

Now, a transformation from one to another coordinate system should be made possible. This is achieved by using transformation matrixes, which should connect three coordinate systems with trihedron $\{\mathbf{e}_{i0}\}$, $\{\mathbf{e}_i\}$ and $\{\mathbf{j}_i\}$, $i = 1,2,3$ to one another. Firstly, the bases $\{\mathbf{j}_i\}$ and $\{\mathbf{e}_i\}$ are considered, in order to create a corresponding transformation matrix. The basis $\{\mathbf{j}_i\}$ can be transformed to the basis $\{\mathbf{e}_i\}$ using three rotations. Logically, this results in three rotations about an axis by angle θ_i; firstly, about the axis $\mathbf{j}_1$ and then about a new axis, resulting from axis $\mathbf{j}_2$. Finally, the rotation about the third axis into the final position. The transformation matrix that results is also named a rotation matrix. This series of rotations results in the following order of the steps for rotational movements, as depicted in Fig. 4.11:

- Rotation about axis $\mathbf{j}_1$ by angle θ_1 with a new position of the trihedron $\{\mathbf{e}_i^1\}$,
- Rotation about axis $\mathbf{e}_2^1$ by angle θ_2 with a new position of the trihedron $\{\mathbf{e}_i^2\}$,
- Rotation about axis $\mathbf{e}_3^2$ by angle θ_3 with a final position, that corresponds to the orientation of the trihedron $\{\mathbf{e}_i\}$.

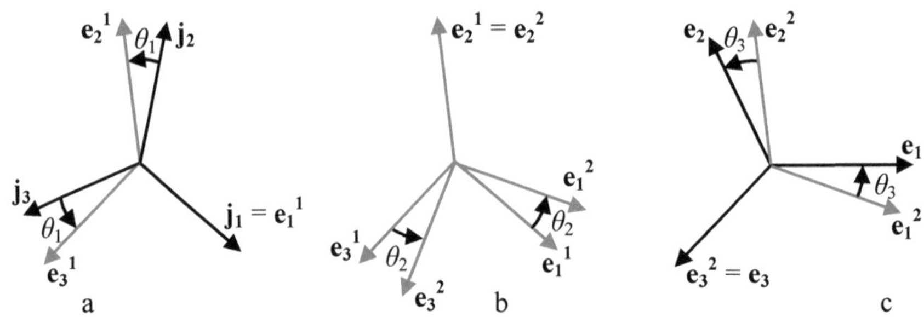

Fig. 4.11: Transformation of the unit vectors $\mathbf{j}_i$ into the unit vectors $\mathbf{e}_i$ using three rotations; a – rotation about axis $\mathbf{j}_1$ by the angle θ_1; b – rotation about axis $\mathbf{e}_2^1$ by the angle θ_2; c – rotation about axis $\mathbf{e}_3^2$ by the angle θ_3

The first step is described through the following relationships between the original unit vectors $\mathbf{j}_i$ and the unit vectors $\mathbf{e}_i^1$:

$$\begin{aligned}
\mathbf{e}_1^1 &= 1\mathbf{j}_1 + 0\mathbf{j}_2 + 0\mathbf{j}_3, \\
\mathbf{e}_2^1 &= 0\mathbf{j}_1 + \cos\theta_1 \mathbf{j}_2 + \sin\theta_1 \mathbf{j}_3, \\
\mathbf{e}_3^1 &= 0\mathbf{j}_1 - \sin\theta_1 \mathbf{j}_2 + \cos\theta_1 \mathbf{j}_3.
\end{aligned} \tag{4.82}$$

This results in a matrix, $\underline{\mathbf{T}}_1$:

$$\underline{\mathbf{T}}_1 = \begin{pmatrix} 1 & 0 & 0 \\ 0 & \cos\theta_1 & \sin\theta_1 \\ 0 & -\sin\theta_1 & \cos\theta_1 \end{pmatrix}, \tag{4.83}$$

which allows for the following transformations:

$$\mathbf{e}_i^1 = \underline{\mathbf{T}}_1^T \mathbf{j}_i, \qquad (4.84)$$
$$\mathbf{j}_i = \underline{\mathbf{T}}_1 \mathbf{e}_i^1.$$

The matrix, given above with index "T", is a transposed matrix; for a rotation matrix, the transposed of one matrix is equal to the inverse of this matrix (see [112]). The expression for $\mathbf{e}_i^1$ can be easily proven, here taking the unit vector $\mathbf{e}_2^1$ as an example:

$$\mathbf{e}_2^1 = \underline{\mathbf{T}}_1^T \mathbf{j}_2 = \begin{pmatrix} 1 & 0 & 0 \\ 0 & \cos\theta_1 & -\sin\theta_1 \\ 0 & \sin\theta_1 & \cos\theta_1 \end{pmatrix} \begin{pmatrix} 0 \\ 1 \\ 0 \end{pmatrix} = \begin{pmatrix} 0 \\ \cos\theta_1 \\ \sin\theta_1 \end{pmatrix}. \qquad (4.85)$$

A comparison with the second equation from (4.82) shows the validity of equation (4.84) for $\mathbf{e}_2^1$. The next rotation by the angle θ_2 converts the unit vectors $\mathbf{e}_i^1$ into the unit vectors $\mathbf{e}_i^2$:

$$\mathbf{e}_1^2 = \cos\theta_2 \mathbf{e}_1^1 + 0\mathbf{e}_2^1 - \sin\theta_2 \mathbf{e}_3^1,$$
$$\mathbf{e}_2^2 = 0\mathbf{e}_1^1 + 1\mathbf{e}_2^1 + 0\mathbf{e}_3^1, \qquad (4.86)$$
$$\mathbf{e}_3^2 = \sin\theta_2 \mathbf{e}_1^1 + 0\mathbf{e}_2^1 + \cos\theta_2 \mathbf{e}_3^1.$$

A second rotation matrix, $\underline{\mathbf{T}}_2$, can thus be written:

$$\underline{\mathbf{T}}_2 = \begin{pmatrix} \cos\theta_2 & 0 & -\sin\theta_2 \\ 0 & 1 & 0 \\ \sin\theta_2 & 0 & \cos\theta_2 \end{pmatrix}. \qquad (4.87)$$

This matrix results in the following transformations:

$$\mathbf{e}_i^2 = \underline{\mathbf{T}}_2^T \mathbf{e}_i^1, \qquad (4.88)$$
$$\mathbf{e}_i^1 = \underline{\mathbf{T}}_2 \mathbf{e}_i^2.$$

The last rotation connects the unit vectors $\mathbf{e}_i^2$ with unit vectors $\mathbf{e}_i$ through angle θ_3:

$$\mathbf{e}_1 = \cos\theta_3 \mathbf{e}_1^2 + \sin\theta_3 \mathbf{e}_2^2 + 0\mathbf{e}_3^2,$$
$$\mathbf{e}_2 = -\sin\theta_3 \mathbf{e}_1^2 + \cos\theta_3 \mathbf{e}_2^2 + 0\mathbf{e}_3^2, \qquad (4.89)$$
$$\mathbf{e}_3 = 0\mathbf{e}_1^2 + 0\mathbf{e}_2^2 + 1\mathbf{e}_3^2.$$

This rotation results in the last matrix, $\underline{\mathbf{T}}_3$:

$$\underline{\mathbf{T}}_3 = \begin{pmatrix} \cos\theta_3 & \sin\theta_3 & 0 \\ -\sin\theta_3 & \cos\theta_3 & 0 \\ 0 & 0 & 1 \end{pmatrix}. \qquad (4.90)$$

This matrix can be used for the following transformations:

$$\mathbf{e}_i = \underline{\mathbf{T}}_3^T \mathbf{e}_i^2, \qquad (4.91)$$
$$\mathbf{e}_i^2 = \underline{\mathbf{T}}_3 \mathbf{e}_i.$$

When these three rotations, described in the equations (4.84), (4.88) and (4.91), are combined, then the transformation from $\{\mathbf{j}_i\}$ to $\{\mathbf{e}_i\}$ can be realized in the following form:

$$\mathbf{j}_\mathrm{i} = \underline{\mathbf{T}}_1 \underline{\mathbf{T}}_2 \underline{\mathbf{T}}_3 \mathbf{e}_\mathrm{i} \tag{4.92}$$

or

$$\begin{aligned} \mathbf{j}_\mathrm{i} &= \underline{\mathbf{T}}\mathbf{e}_\mathrm{i}, \\ \mathbf{e}_\mathrm{i} &= \underline{\mathbf{T}}^\mathrm{T} \mathbf{j}_\mathrm{i}. \end{aligned} \tag{4.93}$$

The following applies for the three resultant rotations described by matrix with elements t_{ik} $(i,k = 1,2,3)$, (Fig. 4.12):

$$\underline{\mathbf{T}} = \underline{\mathbf{T}}_1 \underline{\mathbf{T}}_2 \underline{\mathbf{T}}_3 = \begin{pmatrix} t_{11} & t_{12} & t_{13} \\ t_{21} & t_{22} & t_{23} \\ t_{31} & t_{32} & t_{33} \end{pmatrix}$$

$$= \begin{pmatrix} \cos\theta_2 \cos\theta_3 & \cos\theta_2 \sin\theta_3 & -\sin\theta_2 \\ \begin{matrix} -\cos\theta_1 \sin\theta_3 \\ +\sin\theta_1 \sin\theta_2 \cos\theta_3 \end{matrix} & \begin{matrix} \cos\theta_1 \cos\theta_3 \\ +\sin\theta_1 \sin\theta_2 \sin\theta_3 \end{matrix} & \sin\theta_1 \cos\theta_2 \\ \begin{matrix} \sin\theta_1 \sin\theta_3 \\ +\cos\theta_1 \sin\theta_2 \cos\theta_3 \end{matrix} & \begin{matrix} -\sin\theta_1 \cos\theta_3 \\ +\cos\theta_1 \sin\theta_2 \sin\theta_3 \end{matrix} & \cos\theta_1 \cos\theta_2 \end{pmatrix}. \tag{4.94}$$

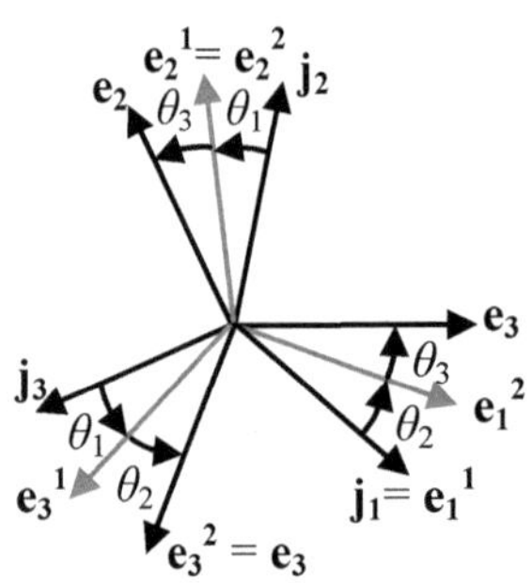

$$\theta_1 \rightarrow \theta_2 \rightarrow \theta_3$$

Fig. 4.12: Representation of the transformation $\{\mathbf{j}_\mathrm{i}\}$ to $\{\mathbf{e}_\mathrm{i}\}$

As well as the transformation of various unit vectors, the transformation matrix allows for a vector $\mathbf{r}_{(\mathrm{j})}$, whose components are defined in the basis $\{\mathbf{j}_\mathrm{i}\}$, to be described through the projections in the basis $\{\mathbf{e}_\mathrm{i}\}$.

$$\mathbf{r}_{(\mathrm{j})} = \begin{pmatrix} r_{(j)1} \\ r_{(j)2} \\ r_{(j)3} \end{pmatrix} - r_{(j)1}\mathbf{j}_1 + r_{(j)2}\mathbf{j}_2 + r_{(j)3}\mathbf{j}_3 \tag{4.95}$$

This is then described in the basis $\{\mathbf{e}_\mathrm{i}\}$ with the designation $\mathbf{r}_{(\mathrm{e})}$, using expression (4.93).

$$\mathbf{r}_{(e)} = r_{(j)1}\underline{\mathbf{T}}\mathbf{e}_1 + r_{(j)2}\underline{\mathbf{T}}\mathbf{e}_2 + r_{(j)3}\underline{\mathbf{T}}\mathbf{e}_3 = r_{(j)1}\begin{pmatrix} t_{11} \\ t_{21} \\ t_{31} \end{pmatrix} + r_{(j)2}\begin{pmatrix} t_{12} \\ t_{22} \\ t_{32} \end{pmatrix} + r_{(j)3}\begin{pmatrix} t_{13} \\ t_{23} \\ t_{33} \end{pmatrix}$$
$$= \begin{pmatrix} r_{(j)1}t_{11} + r_{(j)2}t_{12} + r_{(j)3}t_{13} \\ r_{(j)1}t_{21} + r_{(j)2}t_{22} + r_{(j)3}t_{23} \\ r_{(j)1}t_{31} + r_{(j)2}t_{32} + r_{(j)3}t_{33} \end{pmatrix} = \begin{pmatrix} t_{11} & t_{12} & t_{13} \\ t_{21} & t_{22} & t_{23} \\ t_{31} & t_{32} & t_{33} \end{pmatrix}\begin{pmatrix} r_{(j)1} \\ r_{(j)2} \\ r_{(j)3} \end{pmatrix} = \underline{\mathbf{T}}\mathbf{r}_{(j)} \tag{4.96}$$

Therefore, the following transformation applies, to transfer a vector from one coordinate system into another:

$$\mathbf{r}_{(e)} = \underline{\mathbf{T}}\mathbf{r}_{(j)}\,. \tag{4.97}$$

A difference should be noted between the transformation formula for unit vectors (4.93) and vectors in space, according to equation (4.97). Similarly, a transformation can be written as follows for unit vectors $\mathbf{e}_{i0}$ to unit vectors $\mathbf{e}_i$ ($i = 1,2,3$), as well as for a vector $\mathbf{r}_{(e0)}$ with components in the directions of the basis $\{\mathbf{e}_{i0}\}$ to the vector $\mathbf{r}_{(e)}$ in the directions of the basis $\{\mathbf{e}_i\}$:

$$\mathbf{e}_{i0} = \underline{\mathbf{T}}_{e}\mathbf{e}_i\,, \quad \mathbf{r}_{(e)} = \underline{\mathbf{T}}_{e}\mathbf{r}_{(e0)}\,. \tag{4.98}$$

Using angles θ_{e1}, θ_{e2} und θ_{e3}, a corresponding matrix $\underline{\mathbf{T}}_e$ is formed.

$$\underline{\mathbf{T}}_e = \begin{pmatrix} t_{e11} & t_{e12} & t_{e13} \\ t_{e21} & t_{e22} & t_{e23} \\ t_{e31} & t_{e32} & t_{e33} \end{pmatrix}$$
$$= \begin{pmatrix} \cos\theta_{e2}\cos\theta_{e3} & \cos\theta_{e2}\sin\theta_{e3} & -\sin\theta_{e2} \\ \begin{matrix} -\cos\theta_{e1}\sin\theta_{e3} \\ +\sin\theta_{e1}\sin\theta_{e2}\cos\theta_{e3} \end{matrix} & \begin{matrix} \cos\theta_{e1}\cos\theta_{e3} \\ +\sin\theta_{e1}\sin\theta_{e2}\sin\theta_{e3} \end{matrix} & \sin\theta_{e1}\cos\theta_{e2} \\ \begin{matrix} \sin\theta_{e1}\sin\theta_{e3} \\ +\cos\theta_{e1}\sin\theta_{e2}\cos\theta_{e3} \end{matrix} & \begin{matrix} -\sin\theta_{e1}\cos\theta_{e3} \\ +\cos\theta_{e1}\sin\theta_{e2}\sin\theta_{e3} \end{matrix} & \cos\theta_{e1}\cos\theta_{e2} \end{pmatrix} \tag{4.99}$$

Finally, a transformation is written as follows between unit vectors $\mathbf{e}_{i0}$ and the unit vectors $\mathbf{j}_i$ with $i = 1,2,3$, as well as for vector $\mathbf{r}_{(e0)}$ with components in the directions of the basis $\{\mathbf{e}_{i0}\}$ to the vector $\mathbf{r}_{(j)}$ with directions of the basis $\{\mathbf{j}_i\}$:

$$\begin{aligned} \mathbf{j}_i &= \underline{\mathbf{T}}_0\mathbf{e}_{i0}\,, \\ \mathbf{r}_{(e0)} &= \underline{\mathbf{T}}_0\mathbf{r}_{(j)}\,. \end{aligned} \tag{4.100}$$

Applying the angles θ_{10}, θ_{20} und θ_{30} results in a corresponding matrix $\underline{\mathbf{T}}_0$.

$$\underline{\mathbf{T}}_{\mathbf{0}} = \begin{pmatrix} t_{11,0} & t_{12,0} & t_{13,0} \\ t_{21,0} & t_{22,0} & t_{23,0} \\ t_{31,0} & t_{32,0} & t_{33,0} \end{pmatrix}$$

$$= \begin{pmatrix} \cos\theta_{20}\cos\theta_{30} & \cos\theta_{20}\sin\theta_{30} & -\sin\theta_{20} \\ \begin{matrix} -\cos\theta_{10}\sin\theta_{30} \\ +\sin\theta_{10}\sin\theta_{20}\cos\theta_{30} \end{matrix} & \begin{matrix} \cos\theta_{10}\cos\theta_{30} \\ +\sin\theta_{10}\sin\theta_{20}\sin\theta_{30} \end{matrix} & \sin\theta_{10}\cos\theta_{20} \\ \begin{matrix} \sin\theta_{10}\sin\theta_{30} \\ +\cos\theta_{10}\sin\theta_{20}\cos\theta_{30} \end{matrix} & \begin{matrix} -\sin\theta_{10}\cos\theta_{30} \\ +\cos\theta_{10}\sin\theta_{20}\sin\theta_{30} \end{matrix} & \cos\theta_{10}\cos\theta_{20} \end{pmatrix} \tag{4.101}$$

4.5 Shape of the rod in coordinate systems

Using these transformation matrices, the shape of the deformed rod should be described in a coordinate system, at first in the Cartesian coordinate system, in order to determine the state of the rod in space. Therefore, the vector $\boldsymbol{\kappa}$, which characterize the shape of the rod, should be related to the angles θ_i, $i = 1,2,3$. This relationship is possible by using the equation (4.17) applying the second equation from (4.93).

$$\kappa_{ik} = \frac{d\mathbf{e}_\mathbf{i}}{ds}\mathbf{e}_k = \left(\frac{d\underline{\mathbf{T}}^\mathrm{T}}{ds}\mathbf{j}_\mathbf{i}\right)\left(\underline{\mathbf{T}}^\mathrm{T}\mathbf{j}_k\right) \tag{4.102}$$

The last equation can be written through the components of the transpose of a transformation matrix $\underline{\mathbf{T}}$.

$$\kappa_{ik} = \sum_{n=1}^{3}\frac{dt_{in}^T}{ds}t_{kn}^T \tag{4.103}$$

The following three components of matrix $\underline{\boldsymbol{\kappa}}$ will be written according to the last equation, in order to find the components of the vector $\boldsymbol{\kappa}$:

$$\begin{aligned} \kappa_{23} &= \kappa_1 = \frac{d\theta_1}{ds} - \sin\theta_2\frac{d\theta_3}{ds}, \\ \kappa_{31} &= \kappa_2 = \cos\theta_1\frac{d\theta_2}{ds} + \sin\theta_1\cos\theta_2\frac{d\theta_3}{ds}, \\ \kappa_{12} &= \kappa_3 = -\sin\theta_1\frac{d\theta_2}{ds} + \cos\theta_1\cos\theta_2\frac{d\theta_3}{ds}. \end{aligned} \tag{4.104}$$

These three equations can be given in vector form:

$$\boldsymbol{\kappa} = \underline{\mathbf{T}}_{\boldsymbol{\theta}}\frac{d\boldsymbol{\theta}}{ds} \tag{4.105}$$

with

$$\underline{\mathbf{T}}_{\theta} = \begin{pmatrix} 1 & 0 & -\sin\theta_2 \\ 0 & \cos\theta_1 & \sin\theta_1\cos\theta_2 \\ 0 & -\sin\theta_1 & \cos\theta_1\cos\theta_2 \end{pmatrix}. \qquad (4.106)$$

If equation (4.105) is transformed to obtain the derivative of the angles, the following relationship applies between the vectors $\boldsymbol{\kappa}$ und $\boldsymbol{\theta}$:

$$\frac{d\boldsymbol{\theta}}{ds} = \underline{\mathbf{T}}_{\theta}^{-1}\boldsymbol{\kappa}. \qquad (4.107)$$

This equation relates the shape of the deformed rod and the angles that represent the relationship to the Cartesian coordinate system. An inverse matrix from (4.107) is not the same as a transposed matrix, as is usually the case with rotation matrices. This is written as follows:

$$\underline{\mathbf{T}}_{\theta}^{-1} = \begin{pmatrix} 1 & \sin\theta_1\tan\theta_2 & \cos\theta_1\tan\theta_2 \\ 0 & \cos\theta_1 & -\sin\theta_1 \\ 0 & \sin\theta_1\sec\theta_2 & \cos\theta_1\sec\theta_2 \end{pmatrix}. \qquad (4.108)$$

Considering the transformation (4.98) the equation (4.102) will be rewritten:

$$\kappa_{ik} = \frac{d\mathbf{e}_i}{ds}\mathbf{e}_k = \left(\frac{d\underline{\mathbf{T}}_e^{\mathrm{T}}}{ds}\mathbf{e}_{i0} + \underline{\mathbf{T}}_e^{\mathrm{T}}\frac{d\mathbf{e}_{i0}}{ds}\right)\left(\underline{\mathbf{T}}_e^{\mathrm{T}}\mathbf{e}_{k0}\right). \qquad (4.109)$$

Taking into account the relationship (4.27), which is applied for the directions $\mathbf{e}_{i0}$ and the vector $\boldsymbol{\kappa}_0$, the equation (4.109) has the following form:

$$\kappa_{ik} = \left(\frac{d\underline{\mathbf{T}}_e^{\mathrm{T}}}{ds}\mathbf{e}_{i0} + \underline{\mathbf{T}}_e^{\mathrm{T}}(\boldsymbol{\kappa}_0 \times \mathbf{e}_{i0})\right)\left(\underline{\mathbf{T}}_e^{\mathrm{T}}\mathbf{e}_{k0}\right). \qquad (4.110)$$

Only the elements κ_{23}, κ_{31} and κ_{12} from the expression (4.110) should be taken:

$$\kappa_{23} = \kappa_1 = \kappa_{10} + \frac{d\theta_{e1}}{ds} - \frac{d\theta_{e3}}{ds}\sin\theta_{e2},$$

$$\kappa_{31} = \kappa_2 = \kappa_{20} + \frac{d\theta_{e2}}{ds}\cos\theta_{e1} + \frac{d\theta_{e3}}{ds}\sin\theta_{e1}\cos\theta_{e2}, \qquad (4.111)$$

$$\kappa_{12} = \kappa_3 = \kappa_{30} - \frac{d\theta_{e2}}{ds}\sin\theta_{e1} + \frac{d\theta_{e3}}{ds}\cos\theta_{e1}\cos\theta_{e2}.$$

Summarizing these relationships for κ_1, κ_2 und κ_3 results in the following equation in vector form:

$$\boldsymbol{\kappa} = \boldsymbol{\kappa}_0 + \underline{\mathbf{T}}_{e\theta}\frac{d\boldsymbol{\theta}_e}{ds}. \qquad (4.112)$$

Rewriting equation (4.112) according to the derivative of vector $\boldsymbol{\theta}_e$, results in:

$$\frac{d\boldsymbol{\theta}_e}{ds} = \underline{\mathbf{T}}_{e\theta}^{-1}\left(\boldsymbol{\kappa} - \boldsymbol{\kappa}_0\right), \qquad (4.113)$$

whereby matrix $\underline{\mathbf{T}}_{\mathbf{e\theta}}$ exhibits the following form, as in (4.106):

$$\underline{\mathbf{T}}_{\mathbf{e\theta}} = \begin{pmatrix} 1 & 0 & -\sin\theta_{e2} \\ 0 & \cos\theta_{e1} & \sin\theta_{e1}\cos\theta_{e2} \\ 0 & -\sin\theta_{e1} & \cos\theta_{e1}\cos\theta_{e2} \end{pmatrix}. \quad (4.114)$$

The vector $\boldsymbol{\kappa}_0$ has the following components, that can be determined through a well-known original form of the rod:

$$\boldsymbol{\kappa}_0 = \begin{pmatrix} \kappa_{10} \\ \kappa_{20} \\ \kappa_{30} \end{pmatrix} = \begin{pmatrix} \kappa_{n10} + \dfrac{d\theta_{(10)}}{ds} \\ \kappa_{n30}\sin\theta_{(10)} \\ \kappa_{n30}\cos\theta_{(10)} \end{pmatrix} = \begin{pmatrix} \dfrac{d\theta_{n10}}{ds} + \dfrac{d\theta_{(10)}}{ds} \\ \dfrac{1}{\rho_0}\sin\theta_{(10)} \\ \dfrac{1}{\rho_0}\cos\theta_{(10)} \end{pmatrix}. \quad (4.115)$$

To connect the shape of the rod with the angles between its tangents and coordinate axes, either the equation (4.105) or (4.107) can be used for a Cartesian coordinate system, or the equation (4.112) or (4.113) for a attached coordinate system. These equations are used in an equation system to describe large deflections of a rod.

4.6 Displacement vector

In this section, the displacement of the neutral axis is described. The displacement vector of any point on the neutral axis can be given as the difference between the radius vectors for an original and an already deformed state (Fig. 4.13).

$$\mathbf{u} = \mathbf{r} - \mathbf{r}_0 \quad (4.116)$$

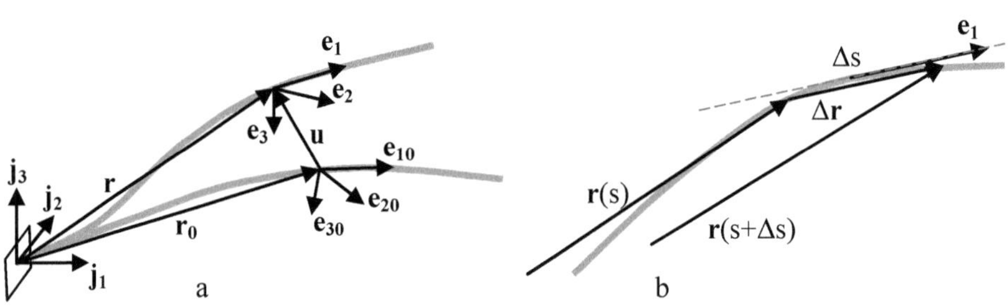

Fig. 4.13: Displacement of the neutral axis; a – representation of the displacement vector **u**; b – representation of the derivative of radius vector **r**

Derivative of this vector equation with respect to parameter s results in a differential equation:

$$\frac{d\mathbf{u}}{ds} = \frac{d\mathbf{r}}{ds} - \frac{d\mathbf{r}_0}{ds}. \quad (4.117)$$

The derivative of a radius vector can be noted as follows, according to the mathematical definition of a derivative:

$$\frac{d\mathbf{r}}{ds} = \lim_{\Delta s \to 0} \frac{\Delta \mathbf{r}}{\Delta s} = \lim_{\Delta s \to 0} \frac{\mathbf{r}(s+\Delta s) - \mathbf{r}(s)}{\Delta s} = \lim_{\Delta s \to 0} \frac{\Delta s}{\Delta s}\mathbf{e}_1 = \mathbf{e}_1 . \tag{4.118}$$

An explanatory representation of this process is given in Fig. 4.13 b. Similarly, the following applies for the derivative of radius vector $\mathbf{r}_0$:

$$\frac{d\mathbf{r}_0}{ds} = \mathbf{e}_{10} . \tag{4.119}$$

Subsequently, equation (4.117) takes on the following simplified form for the displacement of a position s of the rod:

$$\frac{d\mathbf{u}}{ds} = \mathbf{e}_1 - \mathbf{e}_{10} . \tag{4.120}$$

It should be noted that the unit vectors $\mathbf{e}_1$ und $\mathbf{e}_{10}$ are defined in different coordinate systems. Equation (4.120) can be written in the basis $\{\mathbf{e}_{i0}\}$ using the transformation matrices. The position of this trihedron is known, because it is related to the original state of the rod. Therefore, it makes sense to the give the displacement of the rod in the basis $\{\mathbf{e}_{i0}\}$. The vector $\mathbf{u}$ is then split into components in the individual directions:

$$\mathbf{u} = u_1\mathbf{e}_{10} + u_2\mathbf{e}_{20} + u_3\mathbf{e}_{30} . \tag{4.121}$$

As in (4.41), the following applies for the derivative of $\mathbf{u}$ in the basis $\{\mathbf{e}_{i0}\}$:

$$\frac{d\mathbf{u}}{ds} = \frac{d'\mathbf{u}}{ds} + \boldsymbol{\kappa}_0 \times \mathbf{u} . \tag{4.122}$$

With the last expression and the transformation according to (4.98), the following applies for equation (4.120):

$$\frac{d'\mathbf{u}}{ds} = -\boldsymbol{\kappa}_0 \times \mathbf{u} + \left(\underline{\mathbf{T}}_e^{\mathrm{T}} - \underline{\mathbf{E}}\right)\mathbf{e}_{10} . \tag{4.123}$$

Here, $\underline{\mathbf{E}}$ is an identity matrix that has element “1” on its main diagonal, all other elements are “0”. Equation (4.120) can also be represented in the Cartesian coordinate system by using the transposes of the matrices $\underline{\mathbf{T}}$ and $\underline{\mathbf{T}}_0$ or $\underline{\mathbf{T}}_e$ and $\underline{\mathbf{T}}_0$ (see also the description of the equation (4.127)):

$$\begin{aligned} &\frac{d\mathbf{u}}{ds} = \left(\underline{\mathbf{T}}^{\mathrm{T}} - \underline{\mathbf{T}}_0^{\mathrm{T}}\right)\mathbf{j}_1, \\ &\frac{d\mathbf{u}}{ds} - \left(\underline{\mathbf{T}}_0^{\mathrm{T}}\underline{\mathbf{T}}_e^{\mathrm{T}} - \underline{\mathbf{T}}_0^{\mathrm{T}}\right)\mathbf{j}_1 = \mathbf{0}, \end{aligned} \tag{4.124}$$

whereby the displacement vector is also represented in the Cartesian coordinate system:

$$\mathbf{u} = u_x\mathbf{j}_1 + u_y\mathbf{j}_2 + u_z\mathbf{j}_3 . \tag{4.125}$$

Equation (4.124) for a Cartesian coordinate system and equation (4.123) for an attached coordinate system complete the equation system for describing large deflections of a rod in vector form. If a shape of a rod under load should be found in the Cartesian coordinate system, then equation (4.118) can be used in the following form:

$$\frac{d\mathbf{r}}{ds} = \underline{\mathbf{T}}^{\mathrm{T}} \mathbf{j}_1 . \tag{4.126}$$

By using matrices $\underline{\mathbf{T}}_0$ and $\underline{\mathbf{T}}_e$, another form of the equation is obtained, as shown in (4.127). For this, first equation from (4.98) is used to obtain a vector in directions of $\mathbf{e}_{i0}$. Then analogous to the second equation from (4.100) the transformation will be applied to this resulting vector.

$$\frac{d\mathbf{r}}{ds} = \underline{\mathbf{T}}_e^{\mathrm{T}} \mathbf{e}_{10} = \underline{\mathbf{T}}_0^{\mathrm{T}} \underline{\mathbf{T}}_e^{\mathrm{T}} \mathbf{j}_1 \tag{4.127}$$

4.7 Summarizing representation of equations for large deflections

In this section, model equations as non-linear ordinary differential equations are deduced for large deflections of curved, spatial elastic rods. The rods in questions generally have a hollow space that is filled with a pressurized fluid. Furthermore, rods are considered, which have length-constant but limp embedded elements, i.e. a thread or strip. Thus, by using such mathematical model equations, compliant mechanisms, parts of mechanisms and fluid-mechanical actuators can be described. It should be noted when creating such mathematical models and especially when analyzing their results, that the model describes the physical systems as well as possible, as far as the model assumptions and boundary conditions allow.

In order to simplify the use of the model equations given here, they are here below summarized as equation systems. This is achieved according to the nature of the problem posed, whereby these are broken down into two-dimensional and three-dimensional problems. Furthermore, the problems are differentiated from one another by their suitable coordinate systems. Depending on the boundary conditions, the system should be described either in a Cartesian coordinate system or in an attached coordinate system. Alongside vector equations and equations in scalar form for both coordinate systems are summarized for three-dimensional and two-dimensional problems. This results in six model equation systems, which are laid out in Sections 4.7.1-4.7.6. The boundary conditions are written for a cantilever whose clamped end is defined in the Cartesian coordinate origin. Its free end is loaded by external forces and moments.

4.7.1 Vector form of model equations in attached coordinate system in a space

Here the differential equations are summarized as differential equations in vector form in the attached coordinate system. These equations are both equilibrium equations (4.43), (4.46), equation (4.80) that takes material properties into consideration, equation (4.113) for the relationship between the vector $\boldsymbol{\theta}$ and the shape of the deformed rod, which is described using the vector $\boldsymbol{\kappa}$, and the equation (4.123) for the displacement $\mathbf{u}$.

$$\frac{d'\mathbf{Q}}{ds} + \boldsymbol{\kappa} \times (\mathbf{Q} - \mathbf{F_P}) - pA'_P \mathbf{e}_1 + \mathbf{q} = \mathbf{0},$$
$$\frac{d'\mathbf{M}}{ds} + \boldsymbol{\kappa} \times \mathbf{M} + (\mathbf{e}_1 \times \mathbf{Q}) + (\mathbf{h}_T \times \mathbf{q})$$
$$- pA'_P(\mathbf{h} \times \mathbf{e}_1) - F_P(\mathbf{h} \times (\boldsymbol{\kappa} \times \mathbf{e}_1)) + \mathbf{m} = \mathbf{0},$$
$$\frac{d\boldsymbol{\theta}_e}{ds} - \underline{\mathbf{T}}_{e0}^{-1}(\boldsymbol{\kappa} - \boldsymbol{\kappa}_0) = \mathbf{0}, \tag{4.128}$$
$$\frac{d'\mathbf{u}}{ds} + \boldsymbol{\kappa}_0 \times \mathbf{u} - \left(\underline{\mathbf{T}}_e^T - \underline{\mathbf{E}}\right)\mathbf{e}_{10} = \mathbf{0},$$

$$\mathbf{M} = \underline{\mathbf{S}}(\boldsymbol{\kappa} - \boldsymbol{\kappa}_{(0)})$$

This last algebraic equation shows the link between the loads and the resultant deformation, described by $\boldsymbol{\kappa}$. The parameters $\boldsymbol{\kappa}_{(0)}$ and $\boldsymbol{\kappa}_0$ have the components of the same vector, which describe the initial rod shape, but represented in basis $\{\mathbf{e}_i\}$ and $\{\mathbf{e}_{i0}\}$, respectively. Therefore, parameter $\boldsymbol{\kappa}_{(0)}$ can be obtain by following equation:

$$\boldsymbol{\kappa}_{(0)} = \underline{\mathbf{T}}_e \boldsymbol{\kappa}_0 . \tag{4.129}$$

The four differential equations (4.128) should be considered in conjunction with the corresponding boundary conditions. When one end of the rod is clamped and an external force $\mathbf{F}_l$ and a moment $\mathbf{M}_l$ act on the other end, then the following boundary conditions should be considered:

$$\begin{aligned} &\mathbf{Q}(l) = \mathbf{F}_l, \\ &\mathbf{M}(l) = \mathbf{M}_l, \\ &\boldsymbol{\theta}(0) = \mathbf{0}, \\ &\mathbf{u}(0) = \mathbf{0}. \end{aligned} \tag{4.130}$$

If the boundary conditions allow two first and two last equations to be formulated separately, the system of equations can be solved as two separate initial value problems. Otherwise, a boundary value problem is considered, in which four differential equations in vector form can only be solved simultaneously. Using these equations, displacements can be determined in the attached coordinate system. Should a loaded form need to be described in a Cartesian coordinate system, the radius vector can be found using transformation matrices, according to equation (4.127):

$$\frac{d\mathbf{r}}{ds} - \underline{\mathbf{T}}_0^T \underline{\mathbf{T}}_e^T \mathbf{j}_1 = \mathbf{0} . \tag{4.131}$$

The boundary condition for this equation for a rod fixed at point $s = 0$ is:

$$\mathbf{r}(0) = \mathbf{0} . \tag{4.132}$$

Whereby the radius vector has its components in the Cartesian coordinate system.

$$\mathbf{r} = x\mathbf{e}_x + y\mathbf{e}_y + z\mathbf{e}_z \tag{4.133}$$

The components of the matrix $\underline{\mathbf{T}}_0^T$ from equation (4.131) include the angle θ_{i0}, which is determined according to equation (4.107) by using parameter $\boldsymbol{\kappa}_0$, which describes an original shape of the rod and is thus already a known value:

$$\frac{d\boldsymbol{\theta}_0}{ds} = \underline{\mathbf{T}}_{\theta 0}^{-1}\boldsymbol{\kappa}_0 . \qquad (4.134)$$

The inverse matrix used in this case has the same form as the matrix from (4.108):

$$\underline{\mathbf{T}}_{\theta 0}^{-1} = \begin{pmatrix} 1 & \sin\theta_{10}\tan\theta_{20} & \cos\theta_{10}\tan\theta_{20} \\ 0 & \cos\theta_{10} & -\sin\theta_{10} \\ 0 & \sin\theta_{10}\sec\theta_{20} & \cos\theta_{10}\sec\theta_{20} \end{pmatrix} . \qquad (4.135)$$

4.7.2 Scalar form of model equations in attached coordinate system in a space

To determine the scalar equations for an attached coordinate system in a space, the first three differential equations and the last equation of the equation system from (4.128) will be projected in the directions of the basis $\{\mathbf{e}_1, \mathbf{e}_2, \mathbf{e}_3\}$ and the equation for $\mathbf{u}$ in the directions $\mathbf{e}_{10}$, $\mathbf{e}_{20}$ and $\mathbf{e}_{30}$. It is also assumed that the vectors $\mathbf{h}$ and $\mathbf{h}_T$ (Fig. 4.3 a-d) lie on the same line and are described as follows:

$$\mathbf{h} = -h\mathbf{e}_2, \quad \mathbf{h}_T = -h_T\mathbf{e}_2, \qquad h \geq 0, h_T \geq 0 . \qquad (4.136)$$

Due to the above-mentioned simplification, the differential equations in an attached coordinate system in a space to describe the deformation of an elastic rod with a hollow space and an embedded thread or strip can be written as follows:

$$\begin{aligned}
&\frac{dQ_1}{ds} + \kappa_2 Q_3 - \kappa_3 Q_2 - pA_P' + q_1 = 0, \\
&\frac{dQ_2}{ds} - \kappa_1 Q_3 + \kappa_3 (Q_1 - F_P) + q_2 = 0, \\
&\frac{dQ_3}{ds} + \kappa_1 Q_2 - \kappa_2 (Q_1 - F_P) + q_3 = 0, \\
&\frac{dM_1}{ds} + \kappa_2 M_3 - \kappa_3 M_2 - h_T q_3 - hF_P\kappa_2 + m_1 = 0, \\
&\frac{dM_2}{ds} - \kappa_1 M_3 + \kappa_3 M_1 - Q_3 + m_2 = 0, \\
&\frac{dM_3}{ds} + \kappa_1 M_2 - \kappa_2 M_1 + Q_2 + h_T q_1 - hpA_P' + m_3 = 0,
\end{aligned} \qquad (4.137)$$

$$\begin{aligned}
\frac{d\theta_{e1}}{ds} &= (\kappa_1 - \kappa_{10}) + (\kappa_2 - \kappa_{20})\sin\theta_{e1}\tan\theta_{e2} + (\kappa_3 - \kappa_{30})\cos\theta_{e1}\tan\theta_{e2},\\
\frac{d\theta_{e2}}{ds} &= (\kappa_2 - \kappa_{20})\cos\theta_{e1} - (\kappa_3 - \kappa_{30})\sin\theta_{e1},\\
\frac{d\theta_{e3}}{ds} &= (\kappa_2 - \kappa_{20})\sin\theta_{e1}\sec\theta_{e2} + (\kappa_3 - \kappa_{30})\cos\theta_{e1}\sec\theta_{e2},
\end{aligned} \tag{4.138}$$

$$\begin{aligned}
&\frac{du_1}{ds} + \kappa_{20}u_3 - \kappa_{30}u_2 - t_{e11} + 1 = 0,\\
&\frac{du_2}{ds} - \kappa_{10}u_3 + \kappa_{30}u_1 - t_{e12} = 0,\\
&\frac{du_3}{ds} + \kappa_{10}u_2 - \kappa_{20}u_1 - t_{e13} = 0,
\end{aligned} \tag{4.139}$$

$$\begin{aligned}
M_1 &= GI_1(\kappa_1 - \kappa_{(10)}),\\
M_2 &= EI_2(\kappa_2 - \kappa_{(20)}),\\
M_3 &= EI_3(\kappa_3 - \kappa_{(30)}).
\end{aligned} \tag{4.140}$$

Instead of equations (4.138), equations (4.111) can also be used. The parameters t_{eij} with $i,j = 1,2,3$ from equations (4.139), as elements of the transformation matrix $\underline{\mathbf{T}}_{\mathbf{e}}$, can be taken from matrix (4.99). The vector products can be found according to the rules of vector algebra, as is shown below for the following term.

$$\begin{aligned}
\boldsymbol{\kappa}\times(\mathbf{Q}-\mathbf{F_P}) = \boldsymbol{\kappa}\times(\mathbf{Q}-F_P\mathbf{e_1}) = \begin{vmatrix} \mathbf{e_1} & \mathbf{e_2} & \mathbf{e_3}\\ \kappa_1 & \kappa_2 & \kappa_3\\ Q_1 - F_P & Q_2 & Q_3 \end{vmatrix} = \\
(\kappa_2 Q_3 - \kappa_3 Q_2)\mathbf{e_1} - (\kappa_1 Q_3 - \kappa_3(Q_1 - F_P))\mathbf{e_2} + (\kappa_1 Q_2 - \kappa_2(Q_1 - F_P))\mathbf{e_3}
\end{aligned} \tag{4.141}$$

The differential equations from (4.137) for the case of a cantilever loaded by external forces and moments are solved in conjunction with the following boundary conditions:

$$\begin{aligned}
Q_1(l) &= F_{l1}, & M_1(l) &= M_{l1},\\
Q_2(l) &= F_{l2}, & M_2(l) &= M_{l2},\\
Q_3(l) &= F_{l3}, & M_3(l) &= M_{l3}.
\end{aligned} \tag{4.142}$$

For equations (4.138) and (4.139), the following boundary conditions should be taken in consideration in the case of the cantilever:

$$\begin{aligned}
\theta_{e1}(0) &= 0, & u_1(0) &= 0,\\
\theta_{e2}(0) &= 0, & u_2(0) &= 0,\\
\theta_{e3}(0) &= 0, & u_3(0) &= 0.
\end{aligned} \tag{4.143}$$

The algebraic equations from (4.140) should be used for the relationship between M_i and κ_i, and can be applied, for example, to the three equations for the moments as seen in the differential equation system from (4.137).

It should be noted that the parameters $\kappa_{(i0)}$ and κ_{i0} have identically components, but in the various coordinate system.

The parameters κ_{n10} and κ_{n30} are the twist and curvature of the neutral axis of the rod and, along with angle $\theta_{(10)}$, should be known. Therefore, the parameters κ_{i0} can be found. Thus, $\kappa_{(i0)}$ are written by (4.129) as follows:

$$\begin{aligned}\kappa_{(10)} &= \kappa_{10}t_{e11} + \kappa_{20}t_{e12} + \kappa_{30}t_{e13},\\ \kappa_{(20)} &= \kappa_{10}t_{e21} + \kappa_{20}t_{e22} + \kappa_{30}t_{e23},\\ \kappa_{(30)} &= \kappa_{10}t_{e31} + \kappa_{20}t_{e32} + \kappa_{30}t_{e33}.\end{aligned} \tag{4.144}$$

The equation (4.131) can be used to depict the loaded shape of the rod in the Cartesian coordinate system.

$$\begin{aligned}\frac{dx}{ds} &= t_{11,0}t_{e11} + t_{21,0}t_{e12} + t_{31,0}t_{e13},\\ \frac{dy}{ds} &= t_{12,0}t_{e11} + t_{22,0}t_{e12} + t_{32,0}t_{e13},\\ \frac{dz}{ds} &= t_{13,0}t_{e11} + t_{23,0}t_{e12} + t_{33,0}t_{e13}\end{aligned} \tag{4.145}$$

For a rod fixed at one side, the three following boundary conditions are used for these equations:

$$\begin{aligned}x(0) &= 0,\\ y(0) &= 0,\\ z(0) &= 0.\end{aligned} \tag{4.146}$$

4.7.3 Scalar form of model equations in attached coordinate system in a plane

For a plane problem, a number of parameters take a value of zero:

$$\begin{aligned}&Q_3 = M_1 = M_2 = 0,\\ &q_3 = m_1 = m_2 = 0,\\ &\kappa_1 = \kappa_2 = 0,\\ &\theta_1 = \theta_2 = 0,\\ &u_3 = 0.\end{aligned} \tag{4.147}$$

Hence, equations (4.137)–(4.140) are greatly simplified:

$$\begin{aligned}
&\frac{dQ_1}{ds} - \kappa_3 Q_2 - pA'_P + q_1 = 0, \\
&\frac{dQ_2}{ds} + \kappa_3 (Q_1 - F_P) + q_2 = 0, \\
&\frac{dM_3}{ds} + Q_2 + h_T q_1 - hpA'_P + m_3 = 0, \\
&\frac{d\theta_{e3}}{ds} - \kappa_3 + \kappa_{30} = 0, \\
&\frac{du_1}{ds} - \kappa_{30} u_2 - \cos\theta_{e3} + 1 = 0, \\
&\frac{du_2}{ds} + \kappa_{30} u_1 - \sin\theta_{e3} = 0.
\end{aligned} \tag{4.148}$$

Alongside these differential equations, another equation must be taken into consideration when describing the material properties from (4.140), which can be used in the equation for the moment M_3 from (4.148):

$$M_3 = EI_3(\kappa_3 - \kappa_{30}) \, . \tag{4.149}$$

Due to the assumption that this is a plane problem, parameter $\kappa_{(30)}$ is equal to the curvature of the rod in its unloaded state $\kappa_{(30)} = \kappa_{30} = \kappa_{n30}$. The following boundary conditions for the differential equations in (4.148) are considered, for example, for a cantilever loaded by forces and moment.

$$\begin{aligned}
&Q_1(l) = F_{l1}, \qquad && \theta_{e3}(0) = 0, \\
&Q_2(l) = F_{l2}, && u_1(0) = 0, \\
&M_3(l) = M_{l3}, && u_2(0) = 0
\end{aligned} \tag{4.150}$$

If moment M_3 can be eliminated using equation (4.149), a boundary condition for κ_3 can be formulated using this equation:

$$\kappa_3(l) = \frac{M_3(l)}{EI_3} + \kappa_{30}(l) \, . \tag{4.151}$$

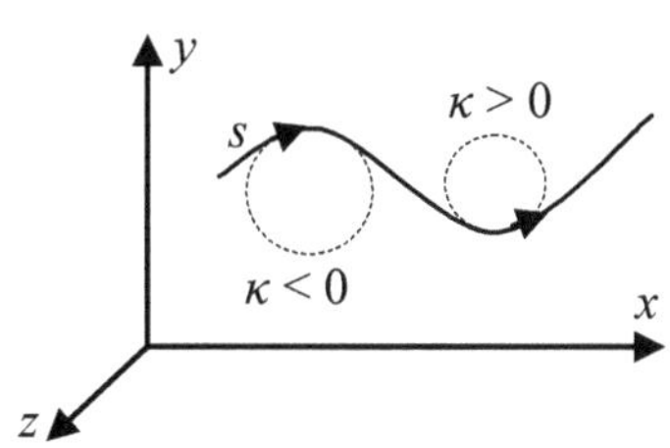

Fig. 4.14: Illustration explaining the algebraic sign of a curve

In plane representations of a curve, its curvature is assigned an algebraic sign ([1]). Such a curvature of the curve is described as positive if a point on the arc element ds rotates positively about the center of curvature in the Cartesian coordinate system as parameter s increases (Fig. 4.14). Should the opposite be the case, the curvature is described as negative.

In order to depict the deformed shaped of the rod in the Cartesian coordinate system, equations based on (4.145) for x and y can be used:

$$\begin{aligned} &\frac{dx}{ds} - \cos(\theta_{e3} + \theta_{30}) = 0, \\ &\frac{dy}{ds} - \sin(\theta_{e3} + \theta_{30}) = 0 \end{aligned} \tag{4.152}$$

with the boundary conditions for a rod whose clamped end is fixed at the Cartesian coordinate origin:

$$\begin{aligned} &x(0) = 0, \\ &y(0) = 0. \end{aligned} \tag{4.153}$$

The parameter θ_{30} for a curved rod, clamped by an angle of θ_0, is found according to equation (4.154).

$$\frac{d\theta_{30}}{ds} - \kappa_{30} = 0, \quad \theta_{30}(0) = \theta_0 \tag{4.154}$$

4.7.4 Vector form of model equations in Cartesian coordinate system in a space

The equilibrium conditions (4.8) and (4.10) should be considered as the base equations for the force and moment equations in the Cartesian coordinate system. In this coordinate system, the forces and moments are split into the directions $\mathbf{j}_i$, $i = 1,2,3$:

$$\begin{aligned} &\mathbf{Q} = Q_x \mathbf{j}_1 + Q_y \mathbf{j}_2 + Q_z \mathbf{j}_3, \\ &\mathbf{M} = M_x \mathbf{j}_1 + M_y \mathbf{j}_2 + M_z \mathbf{j}_3. \end{aligned} \tag{4.155}$$

When a pressure inside the rod must be taken into account, equations for the attached coordinate system should be used. Thus, the load F_P and the parameters h, h_T are not considered: $F_P = 0$, $h = h_T = 0$. In the equation for the moments (4.10), are transformed into a Cartesian coordinate system according to (4.93). Additionally, equations (4.107) or (4.113) and (4.124) must be included in the equations system for describing deflections in the Cartesian coordinate system. Equation (4.80) in consideration of (4.129), which describes the material properties of a rod, completes this equation system. The equation (4.80) was deduced for the components of the moment in the attached coordinate system (Section 4.3) and is assembled by using the transformational matrices for the components M_x, M_y and M_z. The transformation of the moment is carried out using equation (4.97); the differential equations in vector form can thus be written as follows:

$$\frac{d\mathbf{Q}}{ds}+\mathbf{q}=\mathbf{0},$$
$$\frac{d\mathbf{M}}{ds}+(\underline{\mathbf{T}}^{\mathrm{T}}\mathbf{j}_1\times\mathbf{Q})+\mathbf{m}=\mathbf{0},$$
$$\frac{d\boldsymbol{\theta}_e}{ds}-\underline{\mathbf{T}}_{e\theta}^{-1}\left(\boldsymbol{\kappa}-\boldsymbol{\kappa}_0\right)=\mathbf{0}, \tag{4.156}$$
$$\frac{d\mathbf{u}}{ds}-\left(\underline{\mathbf{T}}_0^{\mathrm{T}}\underline{\mathbf{T}}_e^{\mathrm{T}}-\underline{\mathbf{T}}_0^{\mathrm{T}}\right)\mathbf{j}_1=\mathbf{0},$$

$$\underline{\mathbf{T}}\mathbf{M}=\underline{\mathbf{S}}(\boldsymbol{\kappa}-\underline{\mathbf{T}}_e\boldsymbol{\kappa}_0).$$

To find the shape of the deformed rod in Cartesian coordinate system, the equation (4.127) can be used with the corresponding boundary condition, which can be given in the form for a cantilever:

$$\frac{d\mathbf{r}}{ds}-\underline{\mathbf{T}}_0^{\mathrm{T}}\underline{\mathbf{T}}_e^{\mathrm{T}}\mathbf{j}_1=\mathbf{0}, \tag{4.157}$$
$$\mathbf{r}(0)=\mathbf{0}.$$

If the vectors $\boldsymbol{\kappa}_{(0)}$ and $\boldsymbol{\kappa}_0$ are equal, then the last equations can be applied in the shorter form:

$$\frac{d\mathbf{Q}}{ds}+\mathbf{q}=\mathbf{0},$$
$$\frac{d\mathbf{M}}{ds}+(\underline{\mathbf{T}}^{\mathrm{T}}\mathbf{j}_1\times\mathbf{Q})+\mathbf{m}=\mathbf{0},$$
$$\frac{d\boldsymbol{\theta}}{ds}-\underline{\mathbf{T}}_{\theta}^{-1}\boldsymbol{\kappa}=\mathbf{0}, \tag{4.158}$$
$$\frac{d\mathbf{u}}{ds}-\left(\underline{\mathbf{T}}^{\mathrm{T}}-\underline{\mathbf{T}}_0^{\mathrm{T}}\right)\mathbf{j}_1=\mathbf{0},$$

$$\underline{\mathbf{T}}\mathbf{M}=\underline{\mathbf{S}}(\boldsymbol{\kappa}-\boldsymbol{\kappa}_{(0)}).$$

In order to find the moment, the last equation can be rewritten as follows:

$$\mathbf{M}=\underline{\mathbf{T}}^{\mathrm{T}}\underline{\mathbf{S}}(\boldsymbol{\kappa}-\boldsymbol{\kappa}_{(0)})\,. \tag{4.159}$$

Four boundary conditions belong to the differential equations of the equation system in (4.158); taking the example of a cantilever with a force $\mathbf{F}_l$ and moment $\mathbf{M}_l$ acting at the free end, these conditions take on the following form:

$$\mathbf{Q}(l)=\mathbf{F}_l,$$
$$\mathbf{M}(l)=\mathbf{M}_l, \tag{4.160}$$
$$\boldsymbol{\theta}(0)=\mathbf{0},$$
$$\mathbf{u}(0)=\mathbf{0}.$$

If the position of the deformed rod should be found in the Cartesian coordinate system, equation (4.126) can be used with corresponding boundary condition, either instead of or alongside the equation for vector **u**:

$$\begin{aligned}&\frac{d\mathbf{r}}{ds}-\underline{\mathbf{T}}^{\mathrm{T}}\mathbf{j}_1=\mathbf{0},\\&\mathbf{r}(0)=\mathbf{0}.\end{aligned} \tag{4.161}$$

Vector **r** can then be split in its components in the Cartesian coordinate system according to (4.133). The boundary conditions are written for a cantilever with the clamped end at the Cartesian coordinate origin.

4.7.5 Scalar form of model equations in Cartesian coordinate system in a space

The equations from (4.158) for the case $\boldsymbol{\kappa}_{(0)}=\boldsymbol{\kappa}_0$ are now resolved into the directions $\mathbf{j}_i$, $i=1,2,3$ and written in scalar form.

$$\begin{aligned}&\frac{dQ_x}{ds}+q_x=0,\\&\frac{dQ_y}{ds}+q_y=0,\\&\frac{dQ_z}{ds}+q_z=0,\end{aligned} \tag{4.162}$$

$$\begin{aligned}&\frac{dM_x}{ds}+Q_zt_{12}-Q_yt_{13}+m_x=0,\\&\frac{dM_y}{ds}-Q_zt_{11}+Q_xt_{13}+m_y=0,\\&\frac{dM_z}{ds}+Q_yt_{11}-Q_xt_{12}+m_z=0,\end{aligned} \tag{4.163}$$

$$\begin{aligned}&\frac{d\theta_1}{ds}=\kappa_1+\kappa_2\sin\theta_1\tan\theta_2+\kappa_3\cos\theta_1\tan\theta_2,\\&\frac{d\theta_2}{ds}=\kappa_2\cos\theta_1-\kappa_3\sin\theta_1,\\&\frac{d\theta_3}{ds}=\kappa_2\sin\theta_1\sec\theta_2+\kappa_3\cos\theta_1\sec\theta_2,\end{aligned} \tag{4.164}$$

$$\begin{aligned}&\frac{du_x}{ds}-t_{11}+t_{11,0}=0,\\&\frac{du_y}{ds}-t_{12}+t_{12,0}=0,\\&\frac{du_z}{ds}-t_{13}+t_{13,0}=0,\end{aligned} \tag{4.165}$$

$$\begin{aligned} M_x t_{11} + M_y t_{12} + M_z t_{13} &= GI_1(\kappa_1 - \kappa_{(10)}), \\ M_x t_{21} + M_y t_{22} + M_z t_{23} &= EI_2(\kappa_2 - \kappa_{(20)}), \\ M_x t_{31} + M_y t_{32} + M_z t_{33} &= EI_3(\kappa_3 - \kappa_{(30)}) \end{aligned} \tag{4.166}$$

The last three algebraic equations (4.166) can also be written in another form according to (4.159):

$$\begin{aligned} M_x &= t_{11} GI_1(\kappa_1 - \kappa_{(10)}) + t_{21} EI_2(\kappa_2 - \kappa_{(20)}) + t_{31} EI_3(\kappa_3 - \kappa_{(30)}), \\ M_y &= t_{12} GI_1(\kappa_1 - \kappa_{(10)}) + t_{22} EI_2(\kappa_2 - \kappa_{(20)}) + t_{32} EI_3(\kappa_3 - \kappa_{(30)}), \\ M_z &= t_{13} GI_1(\kappa_1 - \kappa_{(10)}) + t_{23} EI_2(\kappa_2 - \kappa_{(20)}) + t_{33} EI_3(\kappa_3 - \kappa_{(30)}). \end{aligned} \tag{4.167}$$

Instead of equation (4.164), the equations from (4.104) can also be used. The equations (4.166), (4.167) and (4.165) include elements of the matrix **T** from (4.94) and the matrix $\mathbf{T_0}$ from (4.101). The following boundary conditions apply for the differential equations (4.162)–(4.165) for a cantilever with a length of l, which is then loaded with acting forces and moments at the free end:

$$\begin{aligned} &Q_x(l) = F_{lx}, && \theta_1(0) = 0, \\ &Q_y(l) = F_{ly}, && \theta_2(0) = 0, \\ &Q_z(l) = F_{lz}, && \theta_3(0) = 0, \\ &M_x(l) = M_{lx}, && u_x(0) = 0, \\ &M_y(l) = M_{ly}, && u_y(0) = 0, \\ &M_z(l) = M_{lz}, && u_z(0) = 0. \end{aligned} \tag{4.168}$$

Alongside the equations for u_x, u_y and u_z, the following differential equations based on the vector equation from (4.161) can be used to determine the position of the deformed rod in Cartesian coordinate system:

$$\begin{aligned} \frac{dx}{ds} - t_{11} &= 0, \\ \frac{dy}{ds} - t_{12} &= 0, \\ \frac{dz}{ds} - t_{13} &= 0. \end{aligned} \tag{4.169}$$

When the components of the transformation matrix t_{1i}, $i = 1,2,3$ from (4.94) are applied to the last three equations, the equations for x, y and z take on the following form:

$$\begin{aligned} \frac{dx}{ds} - \cos\theta_2 \cos\theta_3 &= 0, \\ \frac{dy}{ds} - \cos\theta_2 \sin\theta_3 &= 0, \\ \frac{dz}{ds} + \sin\theta_2 &= 0. \end{aligned} \tag{4.170}$$

The corresponding boundary conditions for a cantilever are as follows:

$$\begin{aligned} &x(0) = 0, \\ &y(0) = 0, \\ &z(0) = 0. \end{aligned} \tag{4.171}$$

4.7.6 Scalar form of model equations in Cartesian coordinate system in a plane

The following parameters are disregarding for a two-dimensional problem:

$$\begin{aligned} &Q_z = M_x = M_y = 0, \\ &q_z = m_x = m_y = 0, \\ &\kappa_1 = \kappa_2 = 0, \\ &\theta_1 = \theta_2 = 0, \\ &u_z = 0. \end{aligned} \tag{4.172}$$

The equations (4.162)–(4.166) can be written while taking into account the assumptions (4.172) made for a plane problem:

$$\begin{aligned} &\frac{dQ_x}{ds} + q_x = 0, \\ &\frac{dQ_y}{ds} + q_y = 0, \\ &\frac{dM_z}{ds} + Q_y \cos\theta_3 - Q_x \sin\theta_3 + m_z = 0, \end{aligned} \tag{4.173}$$

$$\begin{aligned} &\frac{d\theta_3}{ds} - \kappa_3 = 0, \\ &\frac{du_x}{ds} - \cos\theta_3 + \cos\theta_{30} = 0, \\ &\frac{du_y}{ds} - \sin\theta_3 + \sin\theta_{30} = 0, \end{aligned} \tag{4.174}$$

$$M_z = EI_3(\kappa_3 - \kappa_{30}). \tag{4.175}$$

For a plane problem, the parameters $\kappa_{(30)}$ and κ_{30} are the same:

$$\kappa_{30} = \frac{1}{\rho_0}. \tag{4.176}$$

The following boundary conditions for a cantilever with a length l, taking into account the forces and moments acting at the free end, apply for the six differential equations (4.173)–(4.174):

$$\begin{aligned} &Q_x(l) = F_{lx}, && \theta_3(0) = 0, \\ &Q_y(l) = F_{ly}, && u_x(0) = 0, \\ &M_z(l) = M_{lz}, && u_y(0) = 0. \end{aligned} \tag{4.177}$$

To find the position of the deformed rod in Cartesian coordinate system, equations based on (4.170) can be used:

$$\begin{aligned} &\frac{dx}{ds} - \cos\theta_3 = 0, \\ &\frac{dy}{ds} - \sin\theta_3 = 0, \end{aligned} \tag{4.178}$$

where the following expressions for the boundary conditions of a cantilever apply:

$$\begin{aligned} &x(0) = 0, \\ &y(0) = 0. \end{aligned} \tag{4.179}$$

5 Examples of modeling large deflections of curved rod-like structures

In Chapter 4, differential equations useful in the design of models to describe large deflections of curved rods were obtained and summarized for various cases. Based on these differential equations, several compliant systems will be modeled in this chapter and the results of these models subsequently discussed. Modeling equations from Section 4.7 suitable for the tasks will be selected and used to aid in the design of these models. This selection is based on the nature of the load (follower or direction-constant), as well as the nature of the problem (plane or spatial). A problem can only be described as plane when both the original shape of a rod and its loaded shape are located in the same plane.

If follower forces act on a rod, the equations for an attached coordinate system should be used. Conversely, an equation system for components in the Cartesian coordinate system is advisable when solving problems concerning direction-constant forces and moments. In this way, the boundary conditions can be expressed far more simply, and this allows the equations to be solved more easily.

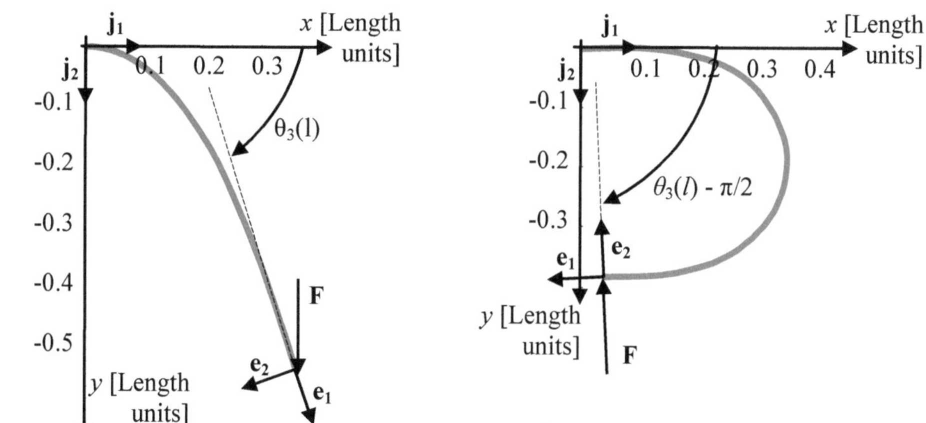

Fig. 5.1: Boundary conditions for the direction-constant and follower loads; a – deformation of a rod loaded with a direction-constant force; b – deformation of a rod loaded with a follower force; the forces in both cases a and b have the same value (compare with Fig. 4.1)

Using the example for the deformation of a rod with a length illustrated in Fig. 5.1, the formulation of the boundary conditions for a follower or direction-constant force can be compared. The direction-constant force (Fig. 5.1 a) can be applied to the boundary conditions for the component forces of the Cartesian coordinate system as follows:

https://doi.org/10.1515/9783110479744-097

$$\begin{aligned} Q_x(l) &= 0, \\ Q_y(l) &= F. \end{aligned} \tag{5.1}$$

In an attached coordinate system, the boundary conditions are written out far more elaborately:

$$\begin{aligned} Q_1(l) &= F \sin\theta_3(l), \\ Q_2(l) &= F \cos\theta_3(l). \end{aligned} \tag{5.2}$$

The angle $\theta_3(l)$ here is still unknown and should be detected using the differential equations. This complicates the solving of these differential equations, as is the case with the boundary conditions (5.1) used in the Cartesian coordinate system. By contrast, a follower force (Fig. 5.1 b) takes on the following form in the attached coordinate system:

$$\begin{aligned} Q_1(l) &= 0, \\ Q_2(l) &= F. \end{aligned} \tag{5.3}$$

However, in the Cartesian coordinate system, this force is incorporated into the boundary conditions in a relatively cumbersome way:

$$\begin{aligned} Q_x(l) &= -F \sin\theta_3(l), \\ Q_y(l) &= F \cos\theta_3(l). \end{aligned} \tag{5.4}$$

Thus, systems under follower loads should be modeled in an attached coordinate system. Conversely, compliant systems loaded with direction-constant forces should be modeled in a Cartesian coordinate system. In this way, the boundary conditions can be used in a simplified form, as in (5.1) or (5.3), thus simplifying and expediting the solving of the differential equations. If both follower and direction-constant forces act on a system, individual boundary conditions in the form of (5.2) or (5.4) should be accepted.

If the external forces do not act on the end of the rod, but on another point of the rod within the length L, this load can be directly incorporated into the differential equations. A force F_L, which acts consistently in the direction $\mathbf{j}_1$ at a point s_0, can be included in a condition as $Q_x(s_0) = F_L$ or in an equation as follows:

$$\frac{dQ_x}{ds} + q_x + \delta_D(s - s_0)F_L = 0\,. \tag{5.5}$$

Here, the following applies for the *DIRAC delta function* $\delta_D(s - s_0)$:

$$\delta_D(s - s_0) = \begin{cases} \infty, & s = s_0, \\ 0, & s \neq s_0. \end{cases} \tag{5.6}$$

Integrating the DIRAC delta function results in a *HEAVISIDE function* with the following properties (see also [112]):

$$\int_{-\infty}^{\infty} \delta_D(s - s_0)ds = H(s - s_0) = \begin{cases} 1, & s \geq s_0, \\ 0, & s < s_0. \end{cases} \tag{5.7}$$

If a force F_L acts on the rod at point s_0, the solution of equation (5.5) considering a condition for q_x and a boundary condition:

$$q_x = 0, \quad Q_x(L) = 0 \tag{5.8}$$

can be written using the HEAVISIDE function:

$$Q_x = F_L\left(1 - H(s - s_0)\right). \tag{5.9}$$

Similarly, a distributed force q_0, which acts not on the entire rod but along a line between points s_1 and s_2, can be given using the HEAVISIDE function for equation (5.5):

$$\frac{dQ_x}{ds} + q_0\left(H(s - s_1) - H(s - s_2)\right) + \delta_D(s - s_0)F_L = 0. \tag{5.10}$$

In this integration, the following property of the HEAVISIDE function should be taken into consideration:

$$\int_{-\infty}^{s} H(s - s_1)ds = (s - s_1)H(s - s_1). \tag{5.11}$$

5.1 Plane problems in the attached coordinate system

The focus of the problems considered in this section is a compliant rod that undergoes deformation under follower loads. A prime example of these being fluid-mechanical actuators with a hollow space and deform when an internal pressure is applied. It should be noted that the radial deformation of the rod cross-section cannot be taken into account using the theory for large rod deflections. This means that the actual bending deflections can be smaller than those calculated using this theory. However, the more such structures are reinforced radially, the better the model obtained by using the theory for large deflections can reproduce the actual deflections.

The following four examples present planar fluid-mechanical actuators and their deformations under the follower loads. These examples include a finger of the gripping device, a pipe with fluid flowing through it, a fluid-mechanical actuator that can be used as a probe and a rod with hollow space and with non-constant cross-sections.

5.1.1 A pneumatically actuated gripping finger

The fingers of a gripping device with a length l have a semicircular inner cross-section and are strengthened on their flat surface with a length-constant but limp strip (Fig. 5.2 a-c). Both fingers of a gripper have an inlet at one end for compressed air and are hermetically sealed at the other. Radial deformations are limited by very thin, rib-like structures. Only one finger has been modeled, which is clamped at one end. A trihedron $\{\mathbf{e}_1, \mathbf{e}_2, \mathbf{e}_3\}$ is attached to the neutral axis of the finger, which runs through the center of the embedded strip. Boundary conditions for the finger will hereafter be discussed. The inner pressure causes only one side of the finger to stretch, whereby the side with the embedded strip remains constant in length. Thus, the finger bends and an object can be gripped.

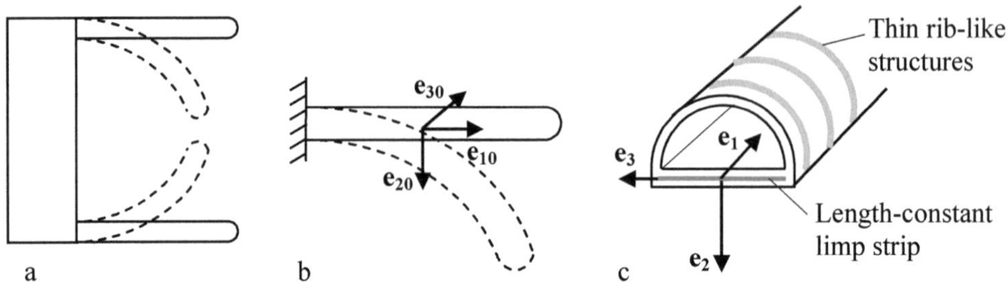

Fig. 5.2: A gripping device with compliant hollow fingers that acts as fluid-mechanical actuators; a – the original and loaded state of the fingers; b – a trihedron $\{\mathbf{e}_{10}, \mathbf{e}_{20}, \mathbf{e}_{30}\}$, attached to the neutral axis of the finger in its unloaded state; c – a semicircular inner cross-section with a length-constant but limp strip and thin rib-like structures to limit radial deformations

The pressure p on the finger's inner front surface can be seen as a resultant force that acts on the centroid of the inner cross-section of the value A_P. With the lever arm h as distance between the centroid of the inner area and the axis $\mathbf{e}_3$, a moment acts about this axis at the end of the finger. A force $\mathbf{F}$ also acts on the end of the finger, due to the effect of the inner pressure on the inner tip of the rod (Fig. 5.3).

$$M = hpA_P, \quad F = pA_P \tag{5.12}$$

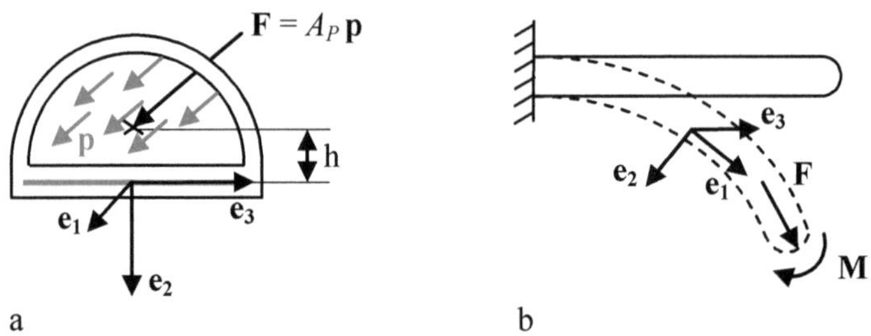

Fig. 5.3: External loads on the finger; a – a resultant force $\mathbf{F}$ of the pressure p on the inner cross-section of the value A_P, which also leads to a moment $\mathbf{M}$; b – finger with the moment $\mathbf{M}$ about axis $\mathbf{e}_3$ and the force $\mathbf{F}$

The equations for these forces, according to the equations from (4.148) are:

$$\begin{aligned} &\frac{dQ_1}{ds} - \kappa_3 Q_2 = 0, \\ &\frac{dQ_2}{ds} + \kappa_3 (Q_1 - pA_P) = 0. \end{aligned} \tag{5.13}$$

The following applies as the boundary conditions:

$$Q_1(l) = F, \quad Q_2(l) = 0. \tag{5.14}$$

A new parameter is introduced to simplify these equations:

$$Q_{1P} = Q_1 - pA_P. \tag{5.15}$$

The equations for the internal forces can thus be written in a shorter form:

$$\frac{dQ_{1P}}{ds} - \kappa_3 Q_2 = 0,$$
$$\frac{dQ_2}{ds} + \kappa_3 Q_{1P} = 0 \tag{5.16}$$

with the new boundary conditions:

$$Q_{1P}(l) = 0, \qquad Q_2(l) = 0. \tag{5.17}$$

Eliminating the curvature from equations in (5.16) results in an equation that only contains forces:

$$Q_{1P}\frac{dQ_{1P}}{ds} + Q_2\frac{dQ_2}{ds} = 0\,. \tag{5.18}$$

After integrating this equation and considering the boundary conditions in (5.17), the following equation can be written:

$$Q_{1P}^2 + Q_2^2 = 0\,. \tag{5.19}$$

It follows from these equations that both internal forces have a value of zero. Thus, an equation for the internal moment M_3 is used from (4.148), which takes on a simple form when considering the results for the internal forces:

$$\frac{dM_3}{ds} = 0\,. \tag{5.20}$$

According to (5.12), a boundary condition is created for this equation

$$M_3(l) = M \tag{5.21}$$

with which a solution for the moment M_3 is found:

$$M_3(s) = hpA_P\,. \tag{5.22}$$

The curvature of the deformed finger can be found using equation (4.149); this curvature remains constant, similar to the moment M_3.

$$\kappa_3(s) = \frac{hpA_P}{EI_3} \tag{5.23}$$

The curvature can be expressed using the curvature radius R of the deformed finger:

$$\frac{1}{R} = \frac{hpA_P}{EI_3}\,. \tag{5.24}$$

This relationship suggests that a larger second moment of area I_3 of the cross-section corresponds with a larger curvature radius of the deformed finger. Furthermore, various problems can be solved using equation (5.24). For example, it is possible to find the inner pressure p required to attain a particular curvature radius, or to find the curvature radius R from a given pressure p. A second moment of area I_3 can also be found through a given pressure and for a desired curvature radius. Should the cross-section of the finger not be constant throughout the hollow space, but instead depend on the coordinate s, the following equations should be solved:

$$\begin{aligned}&\frac{dQ_1}{ds}-\kappa_3 Q_2 - pA_P' = 0,\\ &\frac{dQ_2}{ds}+\kappa_3 (Q_1 - F_P) = 0,\\ &\frac{dM_3}{ds}+Q_2 hpA_P' = 0.\end{aligned} \tag{5.25}$$

The relationship between the internal moment and the curvature should be considered:

$$M_3(s) = EI_3(s)\kappa_3(s)\,. \tag{5.26}$$

The following boundary conditions should also be satisfied:

$$\begin{aligned}&Q_1(l) = pA_P(l), \quad Q_2(l) = 0,\\ &M_3(l) = h(l)pA_P(l).\end{aligned} \tag{5.27}$$

It is advisable to solve these problems numerically (see Section 5.1.4).

5.1.2 A pipe with flowing liquid

A liquid is pumped through a pipe of length L, which flows out orthogonally to the rod axis. Thus a follower force acts on the pipe (Fig. 5.4). The pipe is simplistically modeled as a rod clamped at one end with an effective force acting orthogonally to the rod axis at the other end.

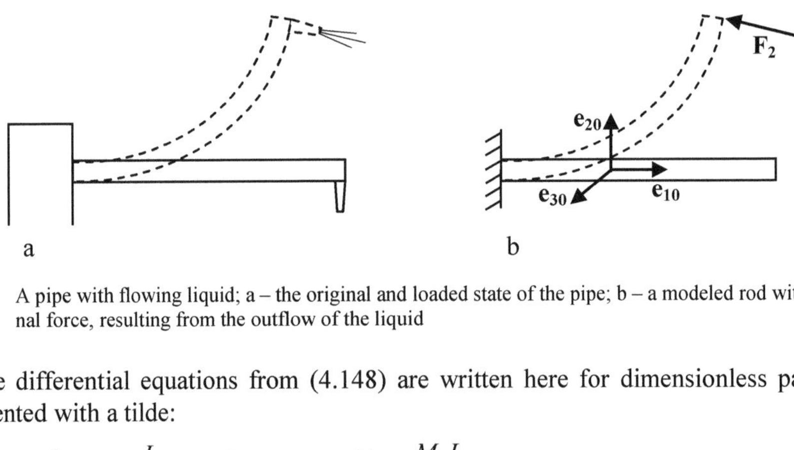

Fig. 5.4: A pipe with flowing liquid; a – the original and loaded state of the pipe; b – a modeled rod with an external force, resulting from the outflow of the liquid

The differential equations from (4.148) are written here for dimensionless parameters, represented with a tilde:

$$\begin{aligned}&\tilde{s} = \frac{s}{L},\quad \tilde{L} = \frac{L}{L} = 1,\quad \tilde{\kappa}_3 = \kappa_3 L,\quad \widetilde{M}_3 = \frac{M_3 L}{EI_3},\\ &\widetilde{F}_2 = \frac{F_2 L^2}{EI_3},\quad \widetilde{EI}_3 = \frac{EI_3}{EI_3} = 1,\quad \widetilde{Q}_i = \frac{Q_i L^2}{EI_3},\quad \tilde{u}_i = \frac{u_i}{L},\ i = 1,2.\end{aligned} \tag{5.28}$$

The dimensionless form of the equations allows the investigations, which are independent from the size and material parameters of the rod. Additionally, the relationship between the moment and the curvature from equation (4.149) is applied to the moment equation from (4.148). This results in six equations, which are used to find the deformation of the pipe:

$$\begin{aligned}
&\frac{d\tilde{Q}_1}{d\tilde{s}} - \tilde{\kappa}_3 \tilde{Q}_2 = 0, \\
&\frac{d\tilde{Q}_2}{d\tilde{s}} + \tilde{\kappa}_3 \tilde{Q}_1 = 0, \\
&\frac{d\tilde{\kappa}_3}{d\tilde{s}} + \tilde{Q}_2 = 0, \\
&\frac{d\theta_{e3}}{d\tilde{s}} - \tilde{\kappa}_3 = 0, \\
&\frac{d\tilde{u}_1}{d\tilde{s}} - \cos\theta_{e3} + 1 = 0, \\
&\frac{d\tilde{u}_2}{d\tilde{s}} - \sin\theta_{e3} = 0.
\end{aligned} \tag{5.29}$$

The third equation of this equation system is rewritten according to Q_2:

$$\tilde{Q}_2 = -\frac{d\tilde{\kappa}_3}{d\tilde{s}}. \tag{5.30}$$

The derivative of this equation is applied to the second equation of the equation system in (5.29). The latter is finally rewritten according to the dimensionless force Q_1:

$$\tilde{Q}_1 = \frac{1}{\tilde{\kappa}_3} \frac{d^2\tilde{\kappa}_3}{d\tilde{s}^2}. \tag{5.31}$$

The last two equations (5.30) and (5.31) are applied to the first equation of the equation system in (5.29) and integrated once accordingly. This results in a non-linear differential equation of the second order for the curvature with an integration constant C_1.

$$\frac{d^2\tilde{\kappa}_3}{d\tilde{s}^2} + \frac{\tilde{\kappa}_3^3}{2} = C_1 \tag{5.32}$$

Boundary conditions are required when calculating the constant C_1. A first condition is formulated for the curvature at the end of the pipe. A second condition can be found from equation (5.31), assuming that no axial force is acting on the end of the pipe.

$$\begin{aligned}
&\tilde{\kappa}_3(1) = 0, \\
&\frac{d^2\tilde{\kappa}_3(1)}{d\tilde{s}^2} = 0
\end{aligned} \tag{5.33}$$

Considering these conditions, the integration constant C_1 has a value of zero. Equation (5.32) can be integrated once more, after it has been multiplied by the derivative of the curvature. This results in an equation with the second integration constant C_2.

$$\left(\frac{d\tilde{\kappa}_3}{d\tilde{s}}\right)^2 + \frac{\tilde{\kappa}_3^4}{4} = C_2 \tag{5.34}$$

Equation (5.30) is used to find the boundary condition at the end of the pipe for the derivative of the curvature:

$$\frac{d\tilde{\kappa}_3(1)}{d\tilde{s}} = -\widetilde{F}_2 . \tag{5.35}$$

Using this condition and the boundary condition for the curvature at the end of the pipe, C_2 is found and then applied to (5.34). This gives the following equation:

$$\left(\frac{d\tilde{\kappa}_3}{d\tilde{s}}\right)^2 + \left(\frac{\tilde{\kappa}_3^2}{2}\right)^2 = \widetilde{F}_2^{\,2} . \tag{5.36}$$

This equation can also be written as:

$$\frac{d\tilde{\kappa}_3}{d\tilde{s}} = \pm\sqrt{\widetilde{F}_2^{\,2} - \left(\frac{\tilde{\kappa}_3^2}{2}\right)^2} . \tag{5.37}$$

The term under radical sign in the last equation should remain positive. This results in a condition for the curvature:

$$\left|\tilde{\kappa}_3\right| \le \sqrt{2\widetilde{F}_2} . \tag{5.38}$$

Thus, the value of the curvature of the deformed pipe is limited by the applied force. Equation (5.37) describes closed curves for various values of the force, based on the relationship between the derivative of the curvature and the curvature itself. The curves (Fig. 5.5a) intersect the coordinate axes at the following points:

$$\left(\pm\sqrt{2\widetilde{F}_2}, 0\right), \quad \left(0, \pm\widetilde{F}_2\right) . \tag{5.39}$$

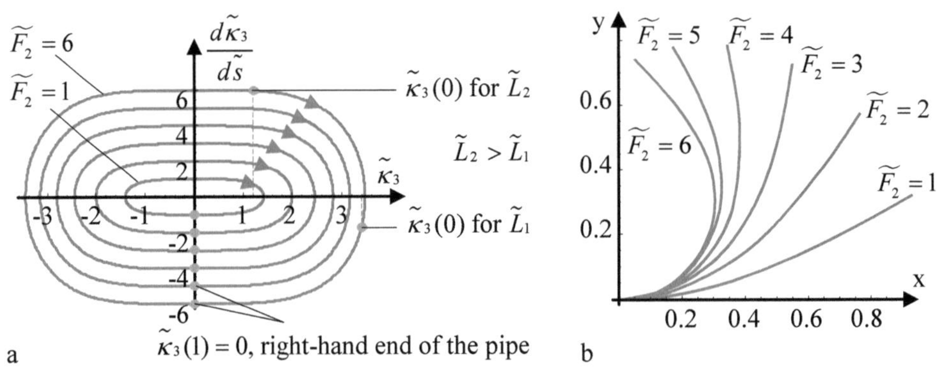

Fig. 5.5: Results for various values of the dimensionless external force; a – lines as a relationship between the curvature and its derivative, a greater length of the pipe corresponds with a greater path on a line; b – six forms of the loaded pipe for forces from 1 to 6 in step of 1

The first condition in (5.33) and the condition in (5.35) for the given positive force result in points on the lines that correspond to the right-hand end of the pipe. The points on the lines for the clamped, left-hand end of the pipe cannot be identified with this depiction, because both the boundary conditions for the curvature and its derivative are unknown. If a line corresponding to a certain external force is considered, a longer pipe corresponds to a longer path on the lines, where the curvature of the pipe never exceeds a maximal value according to the inequality in (5.38).

To find the loaded forms of the pipe under the effects of external forces, two equations for a Cartesian coordinate system can be used instead of the last two equations of the equation system from (5.28):

$$\begin{aligned} &\frac{d\tilde{x}}{d\tilde{s}} - \cos\theta_{e3} = 0, \\ &\frac{d\tilde{y}}{d\tilde{s}} - \sin\theta_{e3} = 0. \end{aligned} \tag{5.40}$$

The angles θ_3 and θ_{e3} are only identical when the unloaded shape of a rod forms a straight line, as in this case. Fig. 5.5 b depicts various pipe shapes for different dimensionless forces from 1 to 6. In addition to the force resulting from the liquid flowing out of the pipe, force acting on the inner front surface of the pipe is also considered. Furthermore, the effect of the flowing liquid on the entire length of the pipe should also be taken into account (Fig. 5.6). A liquid element *dm* with the length of *ds* and with a constant speed of *v* acts with a force $d\mathbf{F}_\mathbf{n}$ orthogonally to the rod axis of the pipe:

$$dF_n\mathbf{e_2} = -dm\frac{v^2}{R}\mathbf{e_2}. \tag{5.41}$$

This force is dependent on the radius of curvature *R* of the deformed pipe. This results in the following relationship for a distributed force on the pipe:

$$q_2 = \frac{dF_n}{ds} = -\rho\pi r^2\frac{v^2}{R} = -\rho\pi\kappa_3 r^2 v^2. \tag{5.42}$$

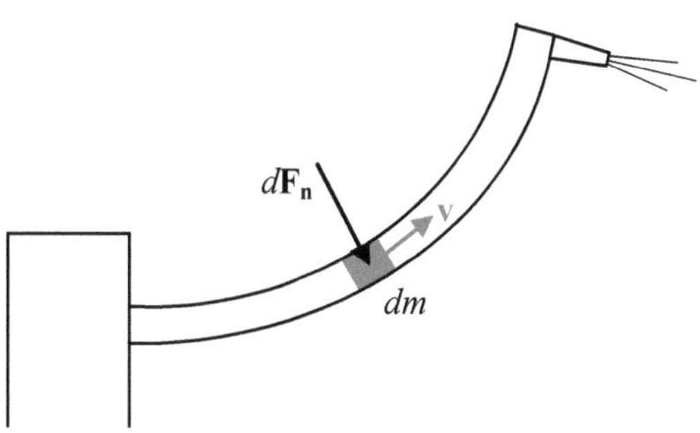

Fig. 5.6: Effect of the flowing liquid on the deformed pipe; a liquid element *dm* with a constant speed of *v* acts with a force $d\mathbf{F}_\mathbf{n}$ orthogonally to the rod axis of the pipe

The distributed force q_2 from (5.42) is applied to the equation from (5.29) for Q_2. The following parameter is introduced for a dimensionless description of the distributed force:

$$\tilde{q}_2 = \frac{q_2 L^3}{EI_3}. \tag{5.43}$$

This results in the following dimensionless equations:

$$\begin{aligned}
&\frac{d\tilde{Q}_1}{d\tilde{s}} - \tilde{\kappa}_3 \tilde{Q}_2 = 0, \\
&\frac{d\tilde{Q}_2}{d\tilde{s}} + \tilde{\kappa}_3 \tilde{Q}_1 + \tilde{q}_2 = 0, \\
&\frac{d\tilde{\kappa}_3}{d\tilde{s}} + \tilde{Q}_2 = 0
\end{aligned} \tag{5.44}$$

with the following boundary conditions:

$$\begin{aligned}
&\tilde{Q}_1(1) = \tilde{F}_1 \\
&\tilde{Q}_2(1) = \tilde{F}_2, \\
&\tilde{\kappa}_3(1) = 0.
\end{aligned} \tag{5.45}$$

These values form is an initial value problem and can be solved numerically. The individual loads; the force from the flowing liquid and the distributed force caused by the liquid flowing through the pipe are proportional to the square of the speed. As this speed increases, the value of the force $\mathbf{F}_2$ also increases. As a result, the curvature of the deformed pipe also increases. An increased curvature increases the value of the distributed force, as described in equation (5.42).

5.1.3 A coated hollow rod

A hollow rod with a length $2l$ and an external, square-shaped cross-section has a coating on the upper-side of the first half of its length. Its lower side is coated along the other half of its length (Fig. 5.7 a). This coating is limp, but remains constant in length. If the pressure inside the hollow space, with a cross-sectional area of A, increases the rod undergoes a deformation. The neutral axis lies on the place of the coating and then passes above and then below the rod. The rod can be separated into two parts with equal lengths. A moment $\mathbf{M}_\mathbf{I}$ and a force $\mathbf{F}_\mathbf{P}$ acts at part I, The effect of the moment $\mathbf{M}_\mathbf{II}$ and the force $\mathbf{F}_\mathbf{P}$ at the end of the rod only applies to part II (Fig. 5.7 b). The moments $\mathbf{M}_\mathbf{I}$ and $\mathbf{M}_\mathbf{II}$, which differ in value due to the asymmetrical position of the hollow space, and the force $\mathbf{F}_\mathbf{P}$ can be depicted as follows, analogous to Section 5.1.1:

$$\begin{aligned}
&M_I = pAh_I, \\
&M_{II} = -pAh_{II}, \\
&F_P = pA.
\end{aligned} \tag{5.46}$$

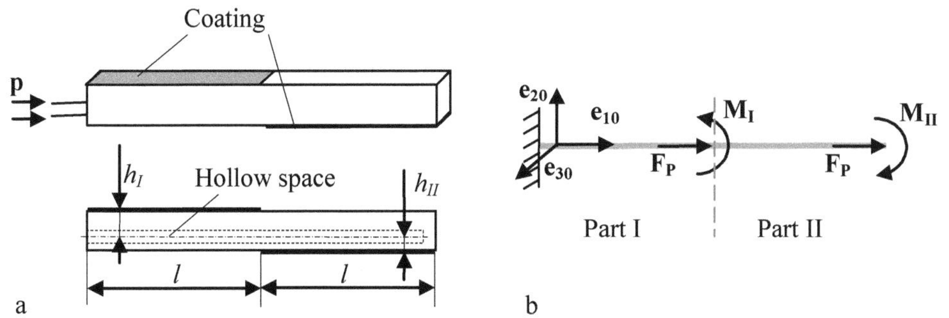

Fig. 5.7: A coated hollow rod; a – a hollow rod with an external, square-shaped cross-section with a limp but length-constant coating; b – modeled rod subjected to two moments and two forces caused by acting of internal pressure

The equilibrium conditions for the forces acting in both parts of the rod can be written as:

$$\frac{dQ_1}{ds} - \kappa_3 Q_2 = 0,$$
$$\frac{dQ_2}{ds} + \kappa_3 (Q_1 - F_P) = 0. \tag{5.47}$$

Introducing a new parameter:

$$Q_{1P} = Q_1 - F_P \tag{5.48}$$

and applying equation (5.47), results in the following equation:

$$Q_{1P}^2 + Q_2^2 = 0\,. \tag{5.49}$$

Taking the boundary conditions into account:

$$Q_{1P}(l) = Q_2(l) = 0,$$
$$Q_{1P}(2l) = Q_2(2l) = 0, \tag{5.50}$$

can aid in solving the internal forces:

$$Q_{1P} = Q_2 = 0\,. \tag{5.51}$$

This is applied in the next equation: the equilibrium condition for the moments. Additionally, the moment at the middle of the rod is introduced into the moment equation using the DIRAC delta function:

$$\frac{dM_3}{ds} + M_I \delta_D (s - l) = 0\,. \tag{5.52}$$

A boundary condition for this equation can be written as:

$$M_3(0) = -M_{II}\,. \tag{5.53}$$

Integrating equation (5.52), while taking into account the boundary condition from (5.53) and the fact that the left half of the rod (part I) is only affected by a moment M_I and the second part only by a moment M_{II}, results in a solution for the internal moment.

$$M_3 = M_I - (M_I - M_{II})H(s-l) \tag{5.54}$$

After introducing the curvature using the relationship from (4.148), this is applied to the equation for θ_{e3}. Furthermore, the equations for each of the two parts are written individually:

$$\frac{d\theta_{e3}}{ds} = \begin{cases} \dfrac{M_I}{EI_{3I}}, & 0 \le s \le l, \\ \dfrac{M_{II}}{EI_{3II}}, & l < s \le 2l. \end{cases} \tag{5.55}$$

The second moment of area $I_{3\mathrm{I}}$ and $I_{3\mathrm{II}}$ are constant in value for each of the parts. The curvature κ_3 also remains constant in each part, so that each part of the rod is shaped like a circular arc. For a further integration of the equation in (5.55), a boundary condition and a transition condition should be considered. This boundary condition is formulated for $s = l$ from the left (indicated with a minus sign) and from the right (indicated with a plus sign).

$$\theta_{e3}(0) = 0, \quad \theta_{e3}(l_-) = \theta_{e3}(l_+) \tag{5.56}$$

The following applies for the angle θ_{e3}:

$$\theta_{e3} = \begin{cases} \dfrac{M_I}{EI_{3I}} s, & 0 \le s \le l, \\ \dfrac{M_{II}}{EI_{3II}} s + \dfrac{M_I l}{EI_{3I}} - \dfrac{M_{II} l}{EI_{3II}}, & l < s \le 2l. \end{cases} \tag{5.57}$$

In order to present the deformed shape of the rod in the Cartesian coordinate system, the equations (4.178) for x and y are used. The angles θ_{e3} and θ_3 equal each other for a straightened, original form of the rod. Two further boundary conditions and required conditions for $x = l$ should be considered when determining the integration constants:

$$\begin{aligned} &x(0) = 0, \quad y(0) = 0, \\ &x(l_-) = x(l_+), \quad y(l_-) = y(l_+). \end{aligned} \tag{5.58}$$

The results for describing the shape of the loaded rod are written as follows:

$$x = \begin{cases} \dfrac{EI_{3I}}{M_I} \sin\left(\dfrac{M_I}{EI_{3I}} s \right), & 0 \le s \le l, \\ \dfrac{EI_{3II}}{M_{II}} \sin\left(\dfrac{M_I l}{EI_{3I}} + \dfrac{M_{II}}{EI_{3II}} (s-l) \right) & \\ + \left(\dfrac{EI_{3I}}{M_I} - \dfrac{EI_{3II}}{M_{II}} \right) \sin \dfrac{M_I l}{EI_{3I}}, & l < s \le 2l. \end{cases} \tag{5.59}$$

$$y=\begin{cases}\dfrac{EI_{3I}}{M_I}\left(1-\cos\left(\dfrac{M_I}{EI_{3I}}s\right)\right), & 0\le s\le l,\\[2ex] -\dfrac{EI_{3II}}{M_{II}}\cos\left(\dfrac{M_I l}{EI_{3I}}+\dfrac{M_{II}}{EI_{3II}}(s-l)\right) & \\[2ex] -\left(\dfrac{EI_{3I}}{M_I}-\dfrac{EI_{3II}}{M_{II}}\right)\cos\dfrac{M_I l}{EI_{3I}}+\dfrac{EI_{3I}}{M_I}, & l<s\le 2l.\end{cases}\tag{5.60}$$

In Fig. 5.8 a a deformed rod depicts with the following values:

$$l=1\text{ unit of the length},\quad \frac{M_I l}{EI_{3I}}=1,\quad \frac{M_{II}l}{EI_{3II}}=-2.\tag{5.61}$$

A relationship between moments can be found, which leads to a special position of the rod. In this position, the end of the rod lies on the x-axis. This relationship results from the condition $y(2l)=0$ from (5.60):

$$\frac{EI_{3II}}{M_{II}}\cos\left(\frac{M_I l}{EI_{3I}}+\frac{M_{II}l}{EI_{3II}}\right)+\left(\frac{EI_{3I}}{M_I}-\frac{EI_{3II}}{M_{II}}\right)\cos\frac{M_I l}{EI_3}-\frac{EI_{3I}}{M_I}=0\,.\tag{5.62}$$

This equation can be solved numerically, choosing a moment, e.g. M_I, and then calculating the moment M_{II}. This equation can have either one or multiple solutions, or indeed no solution, although these solutions must be checked for their usefulness. Fig. 5.8 b depicts a deformed rod, whose end lies on the x-axis. The following values are used, the latter taken from (5.62):

$$l=1\text{ unit of the length},\quad \frac{M_I l}{EI_3}=0.3\,,\quad \frac{M_{II}l}{EI_3}\approx-0.91.\tag{5.63}$$

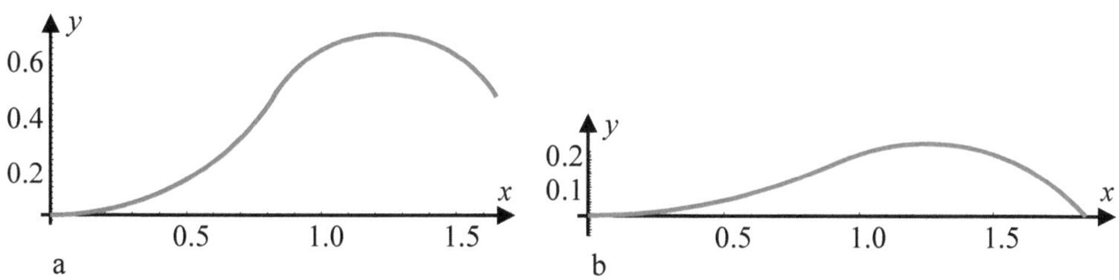

Fig. 5.8: Deformation of the coated, hollow rod; a – a deformed rod with parameters from (5.61); b – a deformed rod with parameters from (5.63), the end of the rod lies on the x-axis

5.1.4 Rods with non-constant cross-sections

This example shows two compliant rods with hollow spaces and non-constant cross-sections along the rod axis (Fig. 5.11). The first rod has a constant radius r_0 and a hollow space with a conical shape. The second rod has a conical outer form and a cylindrical hollow space with a radius r_2. Both rods each contain an embedded fiber, which runs along the rod axis with the same distance to it. If the pressure in the hollow space of one of the rods increases, the rod bends. Such systems are fluid-mechanical actuators, which can be used as probe in mechanics or medical technology. In addition, this method of actuation can also be applied to insert cochlear implants [122].

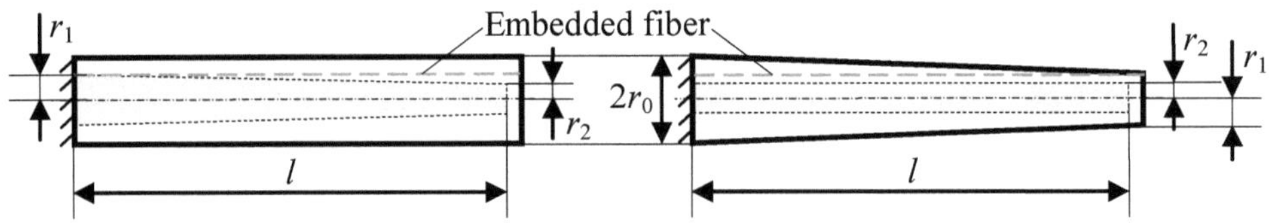

Fig. 5.9: Two rods with hollow spaces and with embedded fibers; a – first rod, a cylindrical rod with a conical inner hollow space; b – second rod, a conical rod with a cylindrical inner hollow space

For the description of the deflection of the rods by inner pressure, the differential equations (4.148) will be used. However, the last two equations are replaced with equations (4.152) to find the deformed shapes of the rods in Cartesian coordinate system.

$$\begin{aligned}
&\frac{dQ_1}{ds} - \kappa_3 Q_2 - pA'_P = 0,\\
&\frac{dQ_2}{ds} + \kappa_3 (Q_1 - pA_P) = 0,\\
&\frac{dM_3}{ds} + Q_2 - hpA'_P = 0,\\
&\frac{d\theta_{e3}}{ds} - \kappa_3 = 0,\\
&\frac{dx}{ds} - \cos\theta_{e3} = 0,\\
&\frac{dy}{ds} - \sin\theta_{e3} = 0
\end{aligned} \tag{5.64}$$

The following parameters apply for the first rod with $r_i(s)$ as inner radius:

$$\begin{aligned}
&r_i(s) = r_1 - \frac{r_1 - r_2}{l} s,\\
&A_P(s) = \pi r_i^2(s),\\
&A'_P(s) = \frac{dA_P}{ds},\\
&I_3(s) = \pi \frac{r_0^4 - r_i^4(s)}{4} + \pi (r_0^2 - r_i^2(s)) r_1^2.
\end{aligned} \tag{5.65}$$

The second rod has an outer radius $r_a(s)$ and the following parameters are applied for the calculation:

$$
\begin{aligned}
&r_a(s) = r_0 - \frac{r_0 - r_1}{l} s, \\
&A_P(s) = \pi r_2^2, \\
&A_P'(s) = 0, \\
&I_3(s) = \pi \frac{r_a^4(s) - r_2^4}{4} + \pi (r_a^2(s) - r_2^2) r_1^2.
\end{aligned} \tag{5.66}
$$

Furthermore, dimensionless parameters are introduced. This allows the investigations to be carried out independently from the size and material parameters of the rods.

$$
\begin{aligned}
&\tilde{s} = \frac{s}{l}, \quad \tilde{l} = \frac{l}{l} = 1, \quad \tilde{\kappa}_3 = \kappa_3 l, \quad \tilde{h} = \frac{h}{l}, \quad \tilde{x} = \frac{x}{l}, \\
&\tilde{y} = \frac{y}{l}, \quad \tilde{r}_j = \frac{r_j}{l}, j = 0,1,2, \quad \tilde{A}_P = \frac{A_P}{l^2}, \quad \widetilde{EI}_3 = \frac{EI_3}{EI_3(s_0)}, \\
&\widetilde{M}_3 = \frac{M_3 l}{EI_3(s_0)}, \quad \tilde{Q}_j = \frac{Q_j l^2}{EI_3(s_0)}, j = 1,2, \quad \tilde{p} = \frac{p l^3}{EI_3(s_0)}
\end{aligned} \tag{5.67}
$$

The parameter $s_0 = l$ is chosen for the first rod and for the second rod the parameter $s_0 = 0$ is applied. The form of the equations (5.64) remains the same also for equations with dimensionless parameters. The following boundary conditions are used for such equations:

$$
\begin{aligned}
&\tilde{Q}_1(l) = \tilde{p}\tilde{A}_P(l), \quad \tilde{Q}_2(l) = 0, \quad \widetilde{M}_3(l) = \tilde{h}\tilde{p}\tilde{A}_P(l), \\
&\tilde{\theta}_3(0) = 0, \quad \tilde{x}(0) = 0, \quad \tilde{y}(0) = 0.
\end{aligned} \tag{5.68}
$$

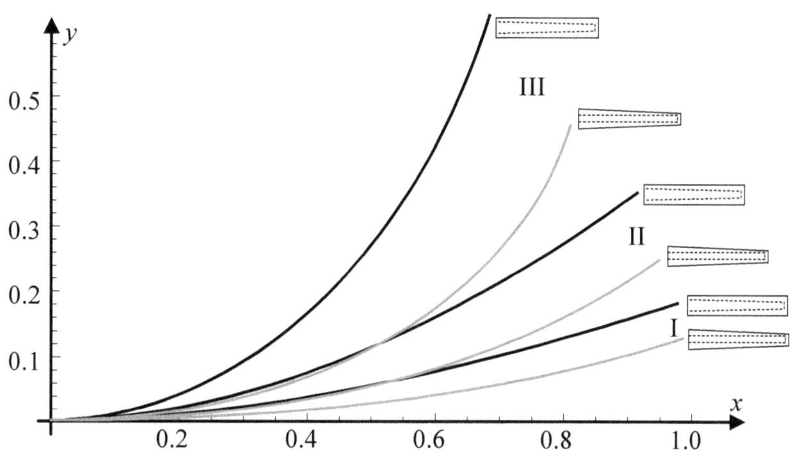

Fig. 5.10: Deformed shapes of two rods in comparison; black curve – first rod, a cylindrical rod with a conical inner hollow space, grey curve – second rod, a conical rod with a cylindrical inner hollow space; I – both rods under dimensionless inner pressure of 200, II – under inner pressure of 400 and III – under inner pressure of 800

The values for the calculation of the deformed rod shape are as follows:

$$
\tilde{r}_0 = 0.1, \quad \tilde{r}_1 = \tilde{h} = 0.08, \quad \tilde{r}_2 = 0.06, \quad \tilde{p} = \{200, 400, 800\}. \tag{5.69}
$$

The results for the deformed rod shapes are shown in Fig. 5.10 for three different values of the inner pressure. The first rod exhibits deflections that are always larger than these of the second rod. It is noteworthy that for all three values of the inner pressure, the angle $\theta_3(l)$ for the second rod remains about 1 % larger than for the first rod in the three given load cases.

5.2 Spatial problems in the attached coordinate system

Spatial problems include problems in describing the deflections of three-dimensional rod structures under the influence of external loads. Also, problems for the description of flat structures that take on a three-dimensional shape during deformation should also be solved using the differential equations for three-dimensional problems. The following examples demonstrate investigations into the deformations of planar and non-planar rod-shaped structures under the influence of direction-constant loads in three-dimensional space. One of these structures is a fluid-mechanical actuator, which can be used as a gripping finger or a probe. Another example is a drill under the influence of external moments.

5.2.1 A helical rod under internal pressure

A rod whose original shape is helical has a hollow space, is radially strengthened and is subjected to internal pressure. A length-constant but limp thread is embedded into the wall of the rod along the rod-axis; this is the neutral axis of the rod. If the pressure inside the hollow space is increased, the helical rod undergoes a deformation. The coordinate origin of the Cartesian coordinate system is placed in the center of the helical form, described through neutral axis, where $z = 0$ also applies for $s = 0$. As z increases, the coordinate s also increases (Fig. 5.11 a-b). The trihedron of the attached coordinate system corresponds to a natural coordinate system of the neutral axis, where the coordinate origin of the attached coordinate system coincides with the embedded thread. This thread has a distance h to the centroid of the cross-section. The shape of the *helix* is characterized by its gradient and the radius R. The gradient is described using an angle ψ between the xy-plane and the tangent to the natural axis, as the embedded thread. Parameter r_i stands for the radius of the inner-space, where r_a is the outer radius (Fig. 5.11 b) of the rod.

In this case, there is a distributed compliance in the rod, although small strains can also lead to large deflections. If the rod is produced from a highly elastic material with non-linear material properties, linear elastic material properties can only be assumed in cases of small rod strains. Parallel axis theorem should be used when finding the second moments of area with respect to the axes of the attached coordinate system:

$$\begin{aligned} I_1 &= \frac{\pi(r_a^4 - r_i^4)}{2} + \pi(r_a^2 - r_i^2)h^2, \\ I_2 &= \frac{\pi(r_a^4 - r_i^4)}{4}, \\ I_3 &= I_2 + \pi(r_a^2 - r_i^2)h^2. \end{aligned} \tag{5.70}$$

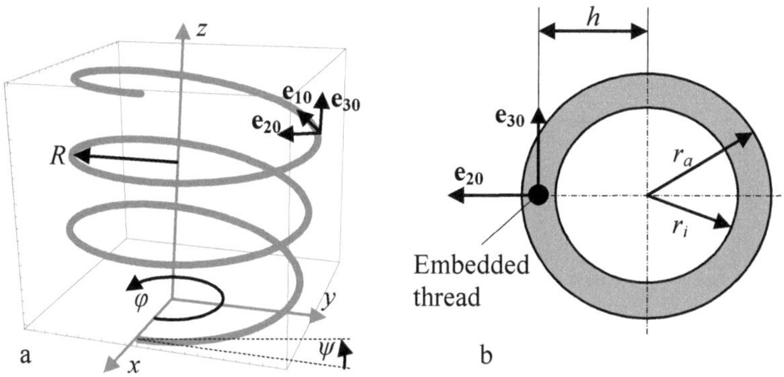

Fig. 5.11: A helical rod with a hollow space and an embedded thread, as a neutral axis, under internal pressure; a – the pitch angle ψ and the radius R characterize the shape of the rod; b – cross-section of the rod with unit vectors $\mathbf{e}_{30}$ and $\mathbf{e}_{20}$, whose origins coincide with the embedded thread; r_i and r_a are the radius of the inner-space and the external radius of the cross-section

A helix can be defined in a Cartesian coordinate system using a parameter φ as a running angle measured from the x-axis to the radius vector in the xy-plane and using the given quantities of R and ψ.

$$\begin{aligned} x &= R\cos\varphi, \\ y &= R\sin\varphi, \\ z &= R\varphi\tan\psi \end{aligned} \qquad (5.71)$$

The attached coordinate system coincides with the natural coordinate system. A representation of the curvature and twist for the natural coordinate system of a helix is given in [112] and is expressed as follows for the helix given in (5.71), whereby both parameters, the curvature and the twist, are constant in value:

$$\begin{aligned} \kappa_{n1} &= \kappa_{10} = \frac{1}{2R}\sin 2\psi, \\ \kappa_{n2} &= \kappa_{20} = 0, \\ \kappa_{n3} &= \kappa_{30} = \frac{1}{R}\cos^2\psi. \end{aligned} \qquad (5.72)$$

For further investigations, dimensionless parameters are used, which are introduced in a similar way to the example from Section 5.1.2. The geometric values relate here, however, to the radius of curvature R to obtain the equations in dimensionless variables:

$$\begin{aligned} &\tilde{s} = \frac{s}{R}, \quad \tilde{L} = \frac{L}{R}, \quad \tilde{R} = \frac{R}{R} = 1, \quad \tilde{h} = \frac{h}{R}, \quad \tilde{r}_i = \frac{r_i}{R}, \quad \tilde{r}_a = \frac{r_a}{R}, \\ &\tilde{F}_P = \frac{F_P R^2}{EI_3}, \quad \tilde{M}_i = \frac{M_i R}{EI_3}, \quad \tilde{G} = \frac{G}{E}, \quad \tilde{E} = \frac{E}{E} = 1, \\ &\tilde{p} = \frac{pR^4}{EI_3}, \quad \tilde{I}_i = \frac{I_i}{I_3}, \quad \tilde{u}_i = \frac{u_i}{R}, \quad \tilde{\kappa}_i = \kappa_i R, \quad i = 1,2,3. \end{aligned} \qquad (5.73)$$

The relationships in (4.140) between the moments and κ_i are now introduced to the equations (4.137) and (4.138) in the following form:

$$\tilde{\kappa}_1 = \frac{\widetilde{M}_1}{\widetilde{G}\widetilde{I}_1} + \tilde{\kappa}_{(10)},$$
$$\tilde{\kappa}_2 = \frac{\widetilde{M}_2}{\widetilde{I}_2} + \tilde{\kappa}_{(20)}, \tag{5.74}$$
$$\tilde{\kappa}_3 = \widetilde{M}_3 + \tilde{\kappa}_{(30)}.$$

Parameters $\kappa_{(i0)}$ are obtained according to (4.144):

$$\tilde{\kappa}_{(10)} = \tilde{\kappa}_{10} t_{e11} + \tilde{\kappa}_{20} t_{e12} + \tilde{\kappa}_{30} t_{e13},$$
$$\tilde{\kappa}_{(20)} = \tilde{\kappa}_{10} t_{e21} + \tilde{\kappa}_{20} t_{e22} + \tilde{\kappa}_{30} t_{e23}, \tag{5.75}$$
$$\tilde{\kappa}_{(30)} = \tilde{\kappa}_{10} t_{e31} + \tilde{\kappa}_{20} t_{e32} + \tilde{\kappa}_{30} t_{e33}.$$

The equations for the internal forces and moments from (4.137) can be written as:

$$\frac{d\widetilde{Q}_1}{d\tilde{s}} + \tilde{\kappa}_2 \widetilde{Q}_3 - \tilde{\kappa}_3 \widetilde{Q}_2 = 0,$$
$$\frac{d\widetilde{Q}_2}{d\tilde{s}} - \tilde{\kappa}_1 \widetilde{Q}_3 + \tilde{\kappa}_3 (\widetilde{Q}_1 - \widetilde{F}_P) = 0,$$
$$\frac{dQ_3}{d\tilde{s}} + \tilde{\kappa}_1 \widetilde{Q}_2 - \tilde{\kappa}_2 (\widetilde{Q}_1 - \widetilde{F}_P) = 0,$$
$$\frac{d\widetilde{M}_1}{d\tilde{s}} + \tilde{\kappa}_2 \widetilde{M}_3 - \tilde{\kappa}_3 \widetilde{M}_2 - \tilde{h}\widetilde{F}_P \tilde{\kappa}_2 = 0, \tag{5.76}$$
$$\frac{dM_2}{d\tilde{s}} - \tilde{\kappa}_1 \widetilde{M}_3 + \tilde{\kappa}_3 \widetilde{M}_1 - \widetilde{Q}_3 = 0,$$
$$\frac{d\widetilde{M}_3}{d\tilde{s}} + \tilde{\kappa}_1 \widetilde{M}_2 - \tilde{\kappa}_2 \widetilde{M}_1 + \widetilde{Q}_2 = 0.$$

For integration of the equations (5.76), the following boundary conditions are taken into account:

$$\widetilde{Q}_1(\tilde{L}) = \widetilde{F}_P, \qquad \widetilde{M}_1(\tilde{L}) = 0,$$
$$\widetilde{Q}_2(\tilde{L}) = 0, \qquad \widetilde{M}_2(\tilde{L}) = 0, \tag{5.77}$$
$$\widetilde{Q}_3(\tilde{L}) = 0, \qquad \widetilde{M}_3(\tilde{L}) = \tilde{h}\widetilde{F}_P.$$

Furthermore, three equations for the derivatives of θ_{ei}, $i = 1,2,3$, can be formulated according to the equations in (4.111):

$$\tilde{\kappa}_1 = \tilde{\kappa}_{10} + \frac{d\theta_{e1}}{d\tilde{s}} - \frac{d\theta_{e3}}{d\tilde{s}} \sin\theta_{e2},$$
$$\tilde{\kappa}_2 = \tilde{\kappa}_{20} + \frac{d\theta_{e2}}{d\tilde{s}} \cos\theta_{e1} + \frac{d\theta_{e3}}{d\tilde{s}} \sin\theta_{e1} \cos\theta_{e2}, \tag{5.78}$$
$$\tilde{\kappa}_3 = \tilde{\kappa}_{30} - \frac{d\theta_{e2}}{d\tilde{s}} \sin\theta_{e1} + \frac{d\theta_{e3}}{d\tilde{s}} \cos\theta_{e1} \cos\theta_{e2}.$$

The following boundary conditions can be used to solve these equations:

$$\theta_{e1}(0) = 0,$$
$$\theta_{e2}(0) = 0, \tag{5.79}$$
$$\theta_{e3}(0) = 0.$$

Depicting the deformations of the rod visually is best achieved in a Cartesian coordinate system. The following applies, based on the scalar equations from (4.145):

$$\frac{d\tilde{x}}{d\tilde{s}} = t_{11,0}t_{e11} + t_{21,0}t_{e12} + t_{31,0}t_{e13},$$
$$\frac{d\tilde{y}}{d\tilde{s}} = t_{12,0}t_{e11} + t_{22,0}t_{e12} + t_{32,0}t_{e13}, \tag{5.80}$$
$$\frac{d\tilde{z}}{d\tilde{s}} = t_{13,0}t_{e11} + t_{23,0}t_{e12} + t_{33,0}t_{e13}.$$

With the boundary conditions:

$$\tilde{x}(0) = \tilde{R},$$
$$\tilde{y}(0) = 0, \tag{5.81}$$
$$\tilde{z}(0) = 0$$

the equation system from (5.80) can be solved numerically. The elements of the transformation matrices t_{eij} and $t_{ij,0}$ must be taken from the matrices in (4.99) and (4.101). To this end, the matrix $\underline{\mathbf{T}}_0$ is necessary, the elements of which include the angles θ_{10}, θ_{20} and θ_{30}. As an example, these are found according to the equations in (4.104), which can be written as three scalar equations:

$$\tilde{\kappa}_{10} = \frac{d\theta_{10}}{d\tilde{s}} - \sin\theta_{20} \frac{d\theta_{30}}{d\tilde{s}},$$
$$\tilde{\kappa}_{20} = \cos\theta_{10} \frac{d\theta_{20}}{d\tilde{s}} + \sin\theta_{10} \cos\theta_{20} \frac{d\theta_{30}}{d\tilde{s}}, \tag{5.82}$$
$$\tilde{\kappa}_{30} = -\sin\theta_{10} \frac{d\theta_{20}}{d\tilde{s}} + \cos\theta_{10} \cos\theta_{20} \frac{d\theta_{30}}{d\tilde{s}}.$$

The three boundary conditions for the last equation system are:

$$\begin{aligned}&\theta_{10}(0)=0,\\&\theta_{20}(0)=0,\\&\theta_{30}(0)=\frac{\pi}{2}.\end{aligned}\tag{5.83}$$

Similar to the equations in (4.170), the original shape of the rod can be described using the following equation system:

$$\begin{aligned}&\frac{d\widetilde{x_0}}{d\tilde{s}}-\cos\theta_{20}\cos\theta_{30}=0,\\&\frac{d\widetilde{y_0}}{d\tilde{s}}-\cos\theta_{20}\sin\theta_{30}=0,\\&\frac{d\tilde{z}_0}{d\tilde{s}}+\sin\theta_{20}=0.\end{aligned}\tag{5.84}$$

The boundary conditions for the last equations are the same as in (5.81). The following material and geometric dimensionless parameters are used to solve the equation system in (5.76), (5.78) and (5.80):

$$\begin{aligned}&\tilde{r}_i=0.05,\quad \tilde{r}_a=0.07,\quad \tilde{h}=0.06,\\&\tilde{L}=6\pi,\quad \psi=0.1,\quad \tilde{G}=\frac{1}{3},\quad \tilde{p}=300.\end{aligned}\tag{5.85}$$

Fig. 5.12 shows the original and loaded shape of the helical rod from two perspectives. The radius of curvature and the entire height of the helical line decrease under internal pressure.

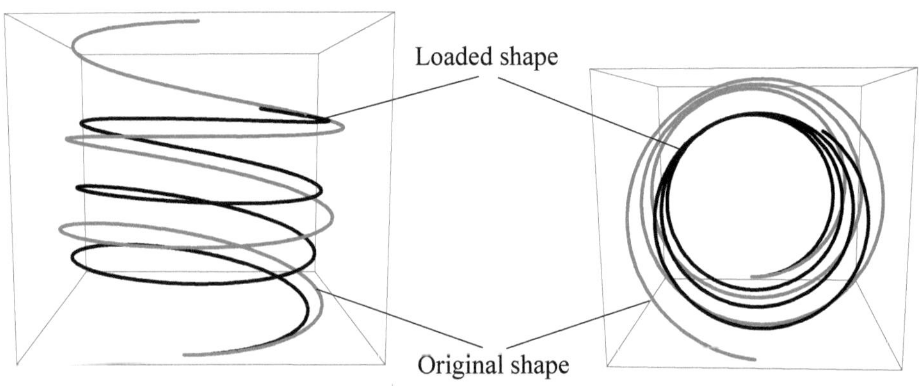

Fig. 5.12: Depiction of the original and loaded shape of the helical rod from two perspectives

If the position of the thread is changed, the deformation of the rod under internal pressure also changes. The change in position of the natural axis is described with the parameter $\theta_{(1)}$ (see equation (4.40)), which represents the original shape of the rod in the following expressions:

$$\tilde{\kappa}_{10} = \frac{1}{2\tilde{R}} \sin 2\psi,$$
$$\tilde{\kappa}_{20} = \frac{1}{\tilde{R}} \cos^2 \psi \sin \theta_{(1)}, \qquad (5.86)$$
$$\tilde{\kappa}_{30} = \frac{1}{\tilde{R}} \cos^2 \psi \cos \theta_{(1)}.$$

In order to depict the deformed rod for different positions of the thread, one winding of the helical line is considered, where the following applies for the length of the rod:

$$\tilde{L} = 2\pi . \qquad (5.87)$$

Fig. 5.13 represents the positions of the thread within the cross-section and the resulting deformed shapes, compared with the original shape of the rod. The displacements of the loaded rod, radius of the curvature and its pitch depend of the thread position in the helical wall.

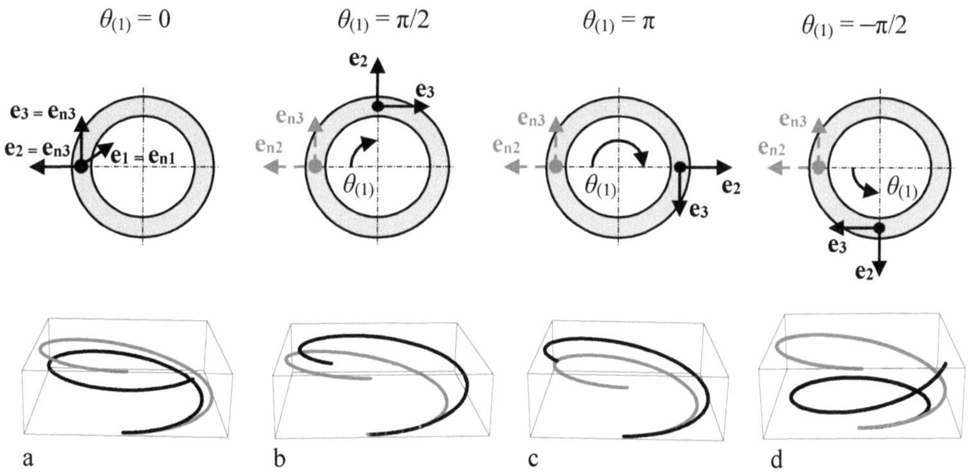

Fig. 5.13: Depiction of the helical rod in its unloaded state and deformed state under internal pressure with different thread positions; a – for $\theta_{(1)} = 0$; b – for $\theta_{(1)} = \pi/2$; c – for $\theta_{(1)} = \pi$; d – for $\theta_{(1)} = -\pi/2$

5.2.2 A drill-bit under load due to a moment

A drill-bit with a length L, which can be described as a rod with a twist in its original state, is acted upon by a moment. This moment consists of three components and acts at the end of the drill-bit.

$$\mathbf{M_L} = M_{L1}\mathbf{e_1} + M_{L2}\mathbf{e_2} + M_{L3}\mathbf{e_3} \qquad (5.88)$$

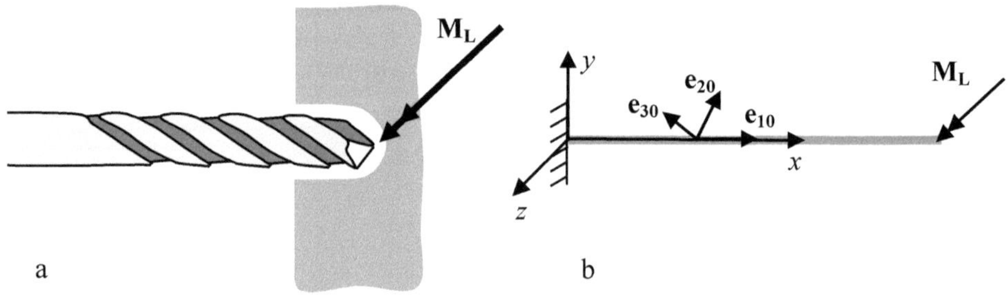

Fig. 5.14: A drill-bit under load due to a moment; a – a drill-bit with a moment acting at its end; b – a model of this drill-bit, the rod has a twist

No external forces act on the rod (the drill-bit), the internal forces thus both have a value of zero. Although the rod is not curved, this is a three-dimensional problem because the rod takes on a three-dimensional form under the effects of the three components of the moment acting upon it. The initial shape of the rod is straight and has approximately rectangular cross-section, values *a* and *b*, and is twisted with four turns about the length *L*.

$$\kappa_{10} = \frac{4\pi}{L}, \quad \kappa_{20} = 0, \quad \kappa_{30} = 0 \tag{5.89}$$

Equations for the moments from (4.137) are then used:

$$\begin{aligned} &\frac{dM_1}{ds} + \kappa_2 M_3 - \kappa_3 M_2 = 0, \\ &\frac{dM_2}{ds} - \kappa_1 M_3 + \kappa_3 M_1 = 0, \\ &\frac{dM_3}{ds} + \kappa_1 M_2 - \kappa_2 M_1 = 0. \end{aligned} \tag{5.90}$$

Considering (4.144), the relationships between the moments and κ_i described in (4.140) are applied:

$$\begin{aligned} &\kappa_1 = \frac{M_1}{GI_1} + \kappa_{10} t_{e11}, \\ &\kappa_2 = \frac{M_2}{EI_2} + \kappa_{10} t_{e21}, \\ &\kappa_3 = \frac{M_3}{EI_2} + \kappa_{10} t_{e31}. \end{aligned} \tag{5.91}$$

The boundary conditions for the equations system in (5.90) are:

$$\begin{aligned} &M_1(L) = M_{L1}, \\ &M_2(L) = M_{L2}, \\ &M_3(L) = M_{L3}. \end{aligned} \tag{5.92}$$

To obtain the parameters θ_{ei} the equations (4.138) can be used with following boundary conditions:

$$\theta_{e1}(L)=0, \quad \theta_{e2}(L)=0, \quad \theta_{e3}(L)=0. \tag{5.93}$$

For the calculation, dimensionless parameters are used, which are introduced as follows:

$$\begin{gathered}
\tilde{s}=\frac{s}{L}, \quad \tilde{L}=\frac{L}{L}=1, \quad \tilde{b}=\frac{b}{L}, \quad \tilde{a}=\frac{a}{L}, \\
\widetilde{M}_i=\frac{M_i R}{E I_3}, \quad \widetilde{G}=\frac{G}{E}, \quad \widetilde{E}=\frac{E}{E}=1, \quad \tilde{I}_i=\frac{I_i}{I_3}, \\
\tilde{x}=\frac{x}{L}, \quad \tilde{y}=\frac{x}{L}, \quad \tilde{y}=\frac{x}{L}, \quad \tilde{\kappa}_i=\kappa_i R, \quad i=1,2,3.
\end{gathered} \tag{5.94}$$

The following material and geometric dimensionless parameters are used to solve the equation system:

$$\tilde{b}=0.1, \quad \tilde{a}=0.05, \quad \widetilde{M}_{Li}=\widetilde{M}_L=0.1, \quad \widetilde{G}=0.38. \tag{5.95}$$

In Fig. 5.15, a drill-bit under load due to different moments is shown. Three different load cases are considered:

$$\begin{aligned}
&\mathrm{I}: \widetilde{\mathbf{M}}_{\mathbf{L}}=\widetilde{M}_L \mathbf{e}_1+\widetilde{M}_L \mathbf{e}_2, \\
&\mathrm{II}: \widetilde{\mathbf{M}}_{\mathbf{L}}=\widetilde{M}_L \mathbf{e}_1+\widetilde{M}_L \mathbf{e}_3, \\
&\mathrm{III}: \widetilde{\mathbf{M}}_{\mathbf{L}}=\widetilde{M}_L \mathbf{e}_1+\widetilde{M}_L \mathbf{e}_2+\widetilde{M}_L \mathbf{e}_3.
\end{aligned} \tag{5.96}$$

The complete displacements of the drill-bit end are 0.018, 0.062 and 0.060 (dimensionless parameters), respectively. The maximum displacement is reached by the drill-bit in case III (Fig. 5.15 b) taking into account the displacements of the entire rod.

When a frictional force acts on a length of the drill from the point s_0 to its end, characterized by a distributed moment m_1, this force can then be incorporated into the first equation of the equation system in (5.90) using the HEAVISIDE function:

$$\frac{dM_1}{ds}+\kappa_2 M_3-\kappa_3 M_2+m_1 H(s-s_0)=0. \tag{5.97}$$

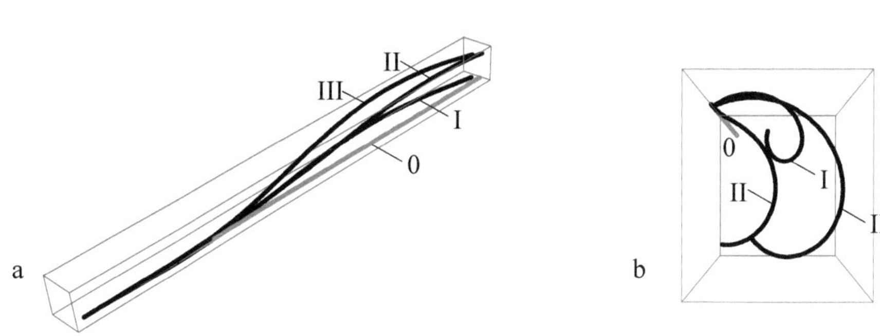

Fig. 5.15: A drill-bit under load due to a moment; a, b – two views of the deformed drill-bit with three different moments: 0 – initial shape, I – under acting of M_{L1} and M_{L2}, II – under acting of M_{L1} and M_{L3}, III – under acting of M_{L1}, M_{L2} and M_{L3} at the end of the drill-bit

However, the boundary condition for the torque moment M_1 remains equal to the condition from (5.92).

5.3 Plane problems in the Cartesian coordinate system

Primarily problems concerning direction-constant loads are solved in the Cartesian coordinate system. Four examples are given in this section that illustrates the use of the differential equations (4.173)-(4.175) for plane problems in the Cartesian coordinate system. Among the examples are compliant sensor elements, a gripping device with a compliant body and a compliant mechanism.

5.3.1 A sensor for measuring dynamic pressures

This first example deals with a compliant element used to measure the dynamic pressure in a pipe filled with a flowing liquid as an element of a sensor system (Fig. 5.16 a). Assuming that the flow speed is equally distributed over the cross-section of the pipe or that the change in speed over the length of the compliant element can be neglected, the distributed load **q** on the length L of the compliant element is presumed to be constant (Fig. 5.16 b).

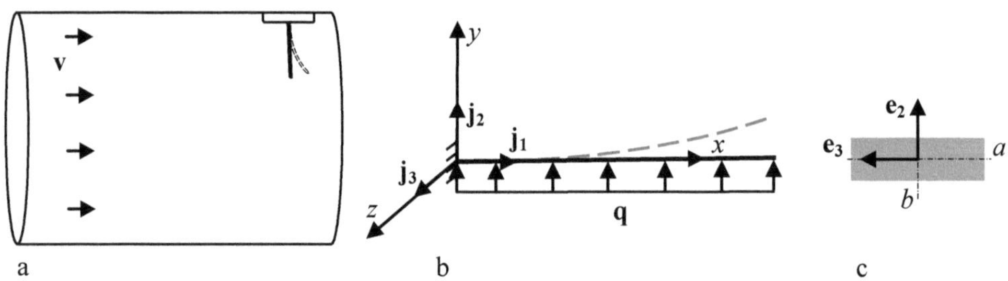

Fig. 5.16: A sensor element for measuring dynamic pressures; a – schematic depiction of a compliant element used to measure the dynamic pressure in a flowing pipe; b – a model of the compliant element, constant distributed force as a direction-constant load; c – cross-section of the compliant element with dimensions b and a

The cross-section of the compliant element is a rectangle with dimensions b and a. The compliant element has a linear shape, where the parameters θ_{30} and $\kappa_{(30)}$ are both zero. The deflection of the compliant rod element is described with the following equations, based on the equation systems (4.173)-(4.175):

$$\begin{aligned}
&\frac{dQ_x}{ds} = 0, \\
&\frac{dQ_y}{ds} + q = 0, \\
&\frac{dM_z}{ds} + Q_y \cos\theta_3 - Q_x \sin\theta_3 = 0, \\
&\frac{d\theta_3}{ds} = \kappa_3, \\
&\frac{du_x}{ds} = \cos\theta_3 - 1, \\
&\frac{du_y}{ds} = \sin\theta_3,
\end{aligned} \tag{5.98}$$

$$M_z = EI_3\kappa_3.$$

For this equation system, corresponding boundary conditions are applied:

$$\begin{aligned}
&Q_x(L) = 0, \qquad \theta_3(0) = 0, \\
&Q_y(L) = 0, \qquad u_x(0) = 0, \\
&M_z(L) = 0, \qquad u_y(0) = 0.
\end{aligned} \tag{5.99}$$

The distributed load, as a distributed force resulting from the liquid flow, is proportional to the square of the speed v, the density of the medium ρ_M in the pipe and the relevant surface area of the rod element.

$$q = k_w L b \rho_M v^2 \tag{5.100}$$

A constant k_w is used here as a proportionality factor. The second moment of area is calculated as follows:

$$I_3 = \frac{ba^3}{12}. \tag{5.101}$$

The first two equations from the equation system in (5.98) give the following solutions for both internal forces:

$$\begin{aligned}
&Q_x = 0, \\
&Q_y = q(L - s).
\end{aligned} \tag{5.102}$$

These are applied to the third equation from the equation system in (5.98), where the relationship between the moment and the curvature (last equation in (5.98)) is also taken in account. This results in four remaining differential equations.

$$\begin{aligned}
&\frac{d\kappa_3}{ds}+\frac{q(L-s)}{EI_3}\cos\theta_3=0,\\
&\frac{d\theta_3}{ds}=\kappa_3,\\
&\frac{du_x}{ds}=\cos\theta_3-1,\\
&\frac{du_y}{ds}=\sin\theta_3
\end{aligned}\tag{5.103}$$

To find a boundary condition for the first equation, for κ_3, the relationship between the moment and the curvature is used:

$$M_z(L)=EI_3\kappa_3(L).\tag{5.104}$$

Thus, the boundary conditions for the equation system in (5.103) can be expressed as follows:

$$\begin{aligned}
&\kappa_z(L)=0,\\
&\theta_3(0)=0,\\
&u_x(0)=0,\\
&u_y(0)=0.
\end{aligned}\tag{5.105}$$

Fig. 5.17 demonstrates a numerical solution to the non-linear equation system in (5.103) with the boundary conditions in (5.105) and the following parameters:

$$\begin{aligned}
&L=20\text{ mm},\quad b=1\text{ mm},\quad a=0.1\text{ mm},\\
&q=0.005\,\frac{\text{N}}{\text{mm}},\quad E=2.1\cdot10^5\,\frac{\text{N}}{\text{mm}^3}.
\end{aligned}\tag{5.106}$$

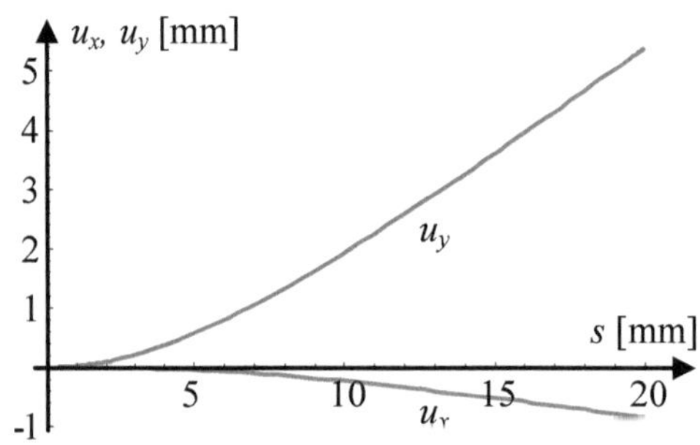

Fig. 5.17: A numerical solution to $u_x(s)$ and $u_y(s)$ for the given values of the parameters from (5.106) for a model of the compliant element in Fig. 5.16

5.3.2 Compliant elements for monitoring angular velocity

Fig. 5.18 demonstrates another example of this kind. Two compliant rod-like elements are attached to a rotating axis using a rigid body lever with a length $2R$. This axis rotates with an angular velocity ω, thus leading to a deformation of the compliant elements with length L.

A system designed in this way can be used to monitor the angular velocity mechanically (Fig. 5.18 a). If there is contact between the compliant element and a contact element, a specific angular velocity is reached or exceeded. Such a compliant element is originally linear in shape. Aerodynamic resistance is disregarded, thus, it can be assumed that the compliant element remains on a plane.

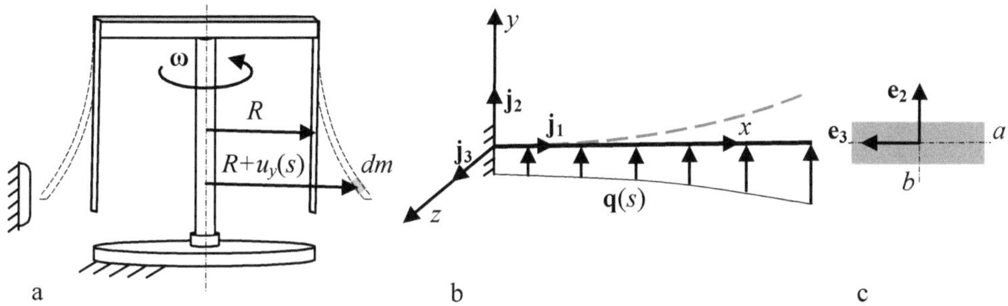

Fig. 5.18: Compliant elements for monitoring angular velocity; a – schematic depiction of a compliant element in a system monitoring angular velocity; b – a model of the compliant element, a distributed force depending on the angular velocity as a direction-constant load; c – cross-section of the compliant element with dimensions *b* and *a*.

A centrifugal force dF acts on a mass dm with corresponding length ds of the compliant rod element. The force depends on the angular velocity and the distance from the axis of rotation.

$$dF = dm\left(R + u_y(s)\right)\omega^2 \tag{5.107}$$

The distributed force on the rod element (Fig. 5.18 b) with a density ρ and cross-sectional dimensions b and a (Fig. 5.18 c) depends on its displacement (deflection on the y axis) and can be written as follows:

$$q(s) = \frac{dF}{ds} = \frac{\rho bads\left(R + u_y(s)\right)\omega^2}{ds} = \rho ba\left(R + u_y(s)\right)\omega^2 . \tag{5.108}$$

This expression is applied to the equation for Q_y, although this equation cannot be integrated analytically. The equation system first takes on the following form, based on the equations from (4.173)–(4.175):

$$\begin{aligned}
&\frac{dQ_x}{ds} = 0, \\
&\frac{dQ_y}{ds} + \rho ba\big(R + k\,u_y(s)\big)\omega^2 = 0, \\
&\frac{dM_z}{ds} + Q_y \cos\theta_3 - Q_x \sin\theta_3 = 0, \\
&\frac{d\theta_3}{ds} = \kappa_3, \qquad\qquad (5.109) \\
&\frac{du_x}{ds} = \cos\theta_3 - 1, \\
&\frac{du_y}{ds} = \sin\theta_3,
\end{aligned}$$

$$M_z = EI_3\kappa_3.$$

Gravity is neglected because it is very small compared to the centrifugal force. Using the parameter k from the second equation, a comparison can be made between the results when taking into account the displacement of the element ($k = 1$) and disregarding its displacement ($k = 0$). The boundary conditions are identical to the boundary conditions (5.99) of the previous example. The first equation gives a solution to Q_x: $Q_x = 0$. Considering the last equation of the equation system in (5.109), the updated equation system has five differential equations:

$$\begin{aligned}
&\frac{dQ_y}{ds} + \rho ba\big(R + k\,u_y(s)\big)\omega^2 = 0, \\
&\frac{d\kappa_3}{ds} + \frac{Q_y}{EI_3}\cos\theta_3 = 0, \\
&\frac{d\theta_3}{ds} = \kappa_3, \qquad\qquad (5.110) \\
&\frac{du_x}{ds} = \cos\theta_3 - 1, \\
&\frac{du_y}{ds} = \sin\theta_3.
\end{aligned}$$

These can be solved when considering the following boundary conditions:

$$\begin{aligned}
&Q_y(L) = 0, \\
&\kappa_z(L) = 0, \\
&\theta_3(0) = 0, \qquad\qquad (5.111) \\
&u_x(0) = 0, \\
&u_y(0) = 0.
\end{aligned}$$

This is a boundary value problem in the context of non-linear differential equations. A corresponding numerical solution for $u_x(s)$ and $u_y(s)$ is given below for the following values of the parameters in Fig. 5.19:

$$\begin{aligned} &\omega = 8\,\frac{\text{rad}}{\text{s}}, \quad L = 40\text{ mm}, \quad b = 5\text{ mm}, \quad a = 0.1\text{ mm}, \\ &R = 10\text{ mm}, \quad \rho = 7.85\cdot 10^{-6}\,\frac{\text{kg}}{\text{mm}^3}, \quad E = 2.1\cdot 10^5\,\frac{\text{N}}{\text{mm}^3}. \end{aligned} \tag{5.112}$$

The solutions for $u_x(s)$ and $u_y(s)$ are visually similar to the corresponding solutions of the last example. A comparison of the solutions for $k = 1$ and $k = 0$ makes it clear that it is necessary in this case to consider the displacement when determining the load.

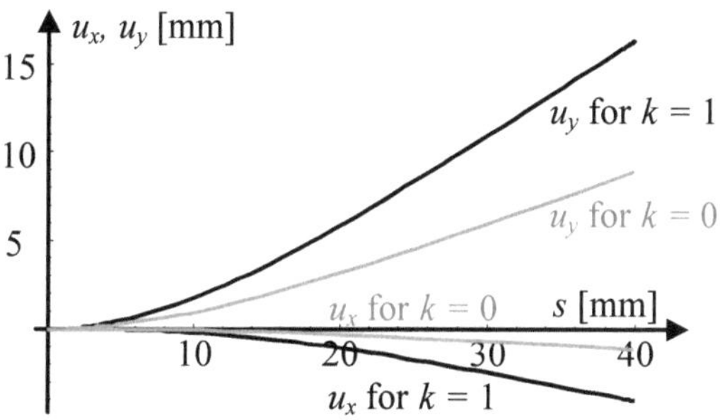

Fig. 5.19: A numerical solution for a model of the compliant element in Fig. 5.18: $k = 1$ and $k = 0$ for the given values of the parameters according to (5.112)

5.3.3 A gripping device with a compliant body

A further example deals with a gripping structure built similarly to the gripping device in [8]. The body of the gripper consists of a compliant ring with a radius R, to which two gripping fingers are attached. An object can therefore be held without energy expenditure in the gripping body, due to its elasticity. If the ring is deformed by an actuator under a force $2\mathbf{F}$, as show in Fig. 5.20 b and d, the gripping fingers move apart. The gripper releases the object and is ready to grip again. Due to its symmetry, only a quarter segment of the compliant ring is modeled. Due to the fact that the tangents at both ends of the segment do not change their direction, one end can be clamped; a second, as depicted in Fig. 5.20 c and e, movably fixed and loaded with a force $\mathbf{F}$. Two given variants for the actuation in Fig. 5.20 b and d should be compared. The length of the compliant element with the shape of a ring segment is:

$$L = \frac{\pi R}{2}. \tag{5.113}$$

Firstly, the case shown in Fig. 5.20 b and c is considered. The curvature of the rod in its unloaded state is positive and reciprocal to the radius of curvature:

$$\kappa_{30} = \frac{1}{R}. \tag{5.114}$$

The first two equations from equation systems (4.173)-(4.175), having been adapted to this situation,

$$\frac{dQ_x}{ds} = 0, \qquad (5.115)$$
$$\frac{dQ_y}{ds} = 0$$

with the boundary conditions

$$Q_x(L) = F, \qquad (5.116)$$
$$Q_y(L) = 0$$

can be solved as follows:

$$Q_x = F, \qquad (5.117)$$
$$Q_y = 0.$$

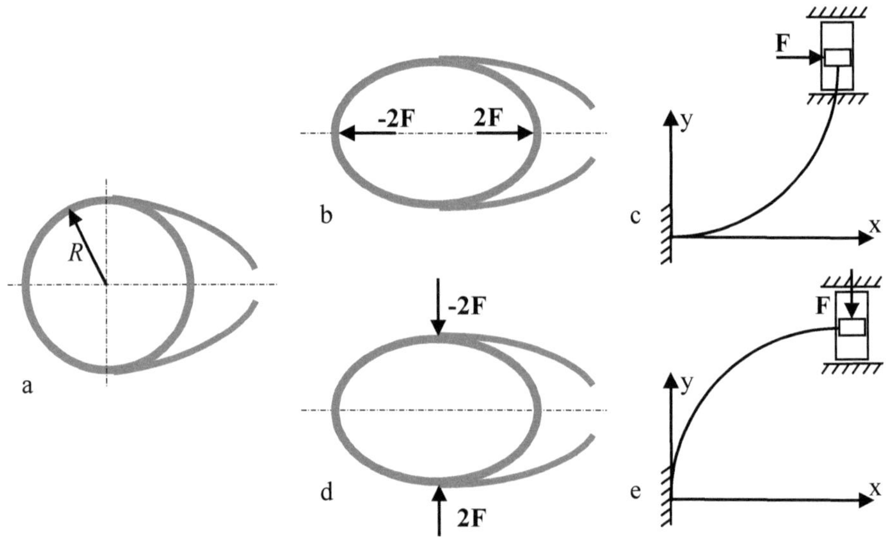

Fig. 5.20: A gripping structure with a compliant ring body; a – original state; b, d – two variants of a force resulting from the actuator; c, e – corresponding mechanical models for both variants, due to the symmetry of the system

Considering these solutions, and the equation from (4.175), only the following differential equations remain:

$$\frac{d\kappa_3}{ds} - \frac{F}{EI_3}\sin\theta_3 = 0,$$
$$\frac{d\theta_3}{ds} = \kappa_3, \qquad (5.118)$$
$$\frac{du_x}{ds} - \cos\theta_3 + \cos\theta_{30} = 0,$$
$$\frac{du_y}{ds} - \sin\theta_3 + \sin\theta_{30} = 0.$$

The angle θ_{30} is found using the following equation from the equation system in (5.118) with following boundary condition:

$$\frac{d\theta_{30}}{ds} = \frac{1}{R}, \quad \theta_{30}(0) = 0. \tag{5.119}$$

Integrating this equation provides the solution for the angle θ_{30}:

$$\theta_{30} = \frac{s}{R}. \tag{5.120}$$

The following boundary conditions for the deformed compliant ring segment can be used:

$$\begin{aligned} &\theta_3(0) = 0, \\ &\theta_3(L) = \frac{\pi}{2}, \\ &u_x(0) = 0, \\ &u_y(0) = 0. \end{aligned} \tag{5.121}$$

The first two equations from the equation system in (5.118) are combined to form a second-order differential equation. This leaves three differential equations (5.122) remaining. Two initial boundary conditions from (5.121) correspond to the first equation of second order from (5.122), and other two boundary conditions can be used for the two first-order equations.

$$\begin{aligned} &\frac{d^2\theta_3}{ds^2} - \frac{F}{EI_3}\sin\theta_3 = 0, \\ &\frac{du_x}{ds} - \cos\theta_3 + \cos\frac{s}{R} = 0, \\ &\frac{du_y}{ds} - \sin\theta_3 + \sin\frac{s}{R} = 0 \end{aligned} \tag{5.122}$$

This equation system can be used to find the angle θ_3, as well as a deflection $\mathbf{u}(s)$ on the plane of deformation of the gripping body. The boundary conditions in (5.121) show that the first equation from (5.122) can be solved independently from the other two equations as a boundary value problem. The two remaining equations represent an initial value problem.

The following problem is now addressed in more detail. When designing such a gripper, the fingers should be fixed to the specific points on the gripping body that allow the fingers to open by the greatest possible angle. Thus, a coordinate s, which corresponds to a maximum change in the angle θ_3 as a positive value of the difference $|\theta_3 - \theta_{30}|$, must be found. In this case, it will suffice to solve only the first equation from (5.122). The equations from (4.178) with boundary conditions from (4.179) are used to depict the deformed shape of the ring segment in the Cartesian coordinate system. In order to reach general conclusions from the results, which are quantitatively dependent neither on the geometric nor material parameters, dimensionless parameters are introduced:

$$\tilde{s} = \frac{s}{L}, \quad \tilde{L} = \frac{L}{L} = 1, \quad \tilde{x} = \frac{x}{L}, \quad \tilde{y} = \frac{y}{L}, \quad \tilde{F} = \frac{FL^2}{EI_3}. \tag{5.123}$$

Applying these parameters to the first equation from (5.122) and the equations from (4.178) results in three dimensionless differential equations:

$$\frac{d^2\theta_3}{d\tilde{s}^2} - \tilde{F}\sin\theta_3 = 0,$$
$$\frac{d\tilde{x}}{d\tilde{s}} - \cos\theta_3 = 0, \tag{5.124}$$
$$\frac{d\tilde{y}}{d\tilde{s}} - \sin\theta_3 = 0$$

with corresponding boundary conditions:

$$\theta_3(0) = 0,$$
$$\theta_3(1) = \frac{\pi}{2}, \tag{5.125}$$
$$\tilde{x}(0) = 0,$$
$$\tilde{y}(0) = 0.$$

Fig. 5.21 shows the solutions for the change in angle θ_3, as well as the shape of a ring segment of the gripping body both before and after a deformation enacted by a force in the Cartesian coordinate system. The results are calculated for two different dimensionless forces with values of 1 and 10. The values from 1 to 10 are assumed to be all applicable actuating forces. These two forces correspond to two maximum values for the change in angle on the ring segment: the points 0.57 and 0.62 (rounded numbers). Assuming that these values, which each give a maximum value of the difference $|\theta_3 - \theta_{30}|$, monotonously increase as the force increases from 1 to 10, the statement below can be made. A suitable interval for the attachment of the fingers on the ring segment of the gripping body is as follows:

$$0.57 \leq \tilde{s}_M \leq 0.62\,. \tag{5.126}$$

The difference $|\theta_3 - \theta_{30}|$ has a value of approximately 0.7 rad (around 40°) for the point at 0.62. The length of the neutral axis of the ring segment corresponds to a dimensionless value of 1, the fingers can thus be attached as show in Fig. 5.23 a.

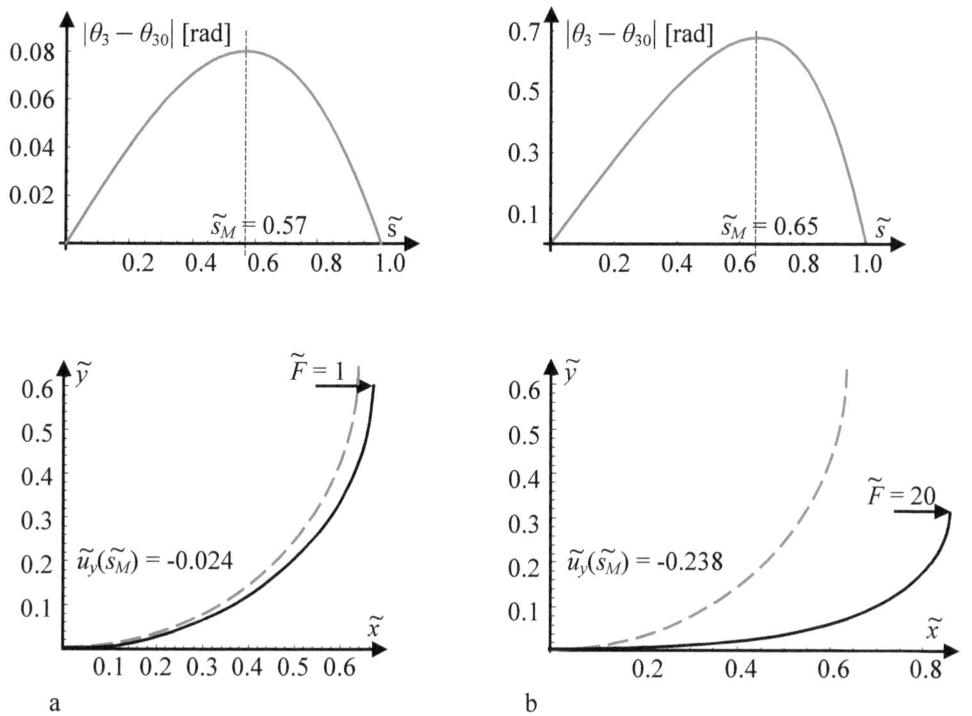

Fig. 5.21: Difference $|\theta_3 - \theta_{30}|$ and the shapes of the ring segment before deformation (dotted line) and after deformation for the rod model from Fig. 5.20 c; a – for a dimensionless force with a value of 1; b – for a dimensionless force with a value of 20

The second variant of the acting force, shows in Fig. 5.20 d, is now considered. The following equations and boundary conditions for this variant can be written in a similar way to the first variant of the acting force:

$$\begin{aligned}
&\frac{d^2\theta_3}{d\tilde{s}^2} + \tilde{F}\cos\theta_3 = 0,\\
&\frac{d\tilde{x}}{d\tilde{s}} - \cos\theta_3 = 0, \qquad (5.127)\\
&\frac{d\tilde{y}}{d\tilde{s}} - \sin\theta_3 = 0,
\end{aligned}$$

$$\begin{aligned}
&\theta_3(0) = \frac{\pi}{2},\\
&\theta_3(1) = 0, \qquad (5.128)\\
&\tilde{x}(0) = 0,\\
&\tilde{y}(0) = 0.
\end{aligned}$$

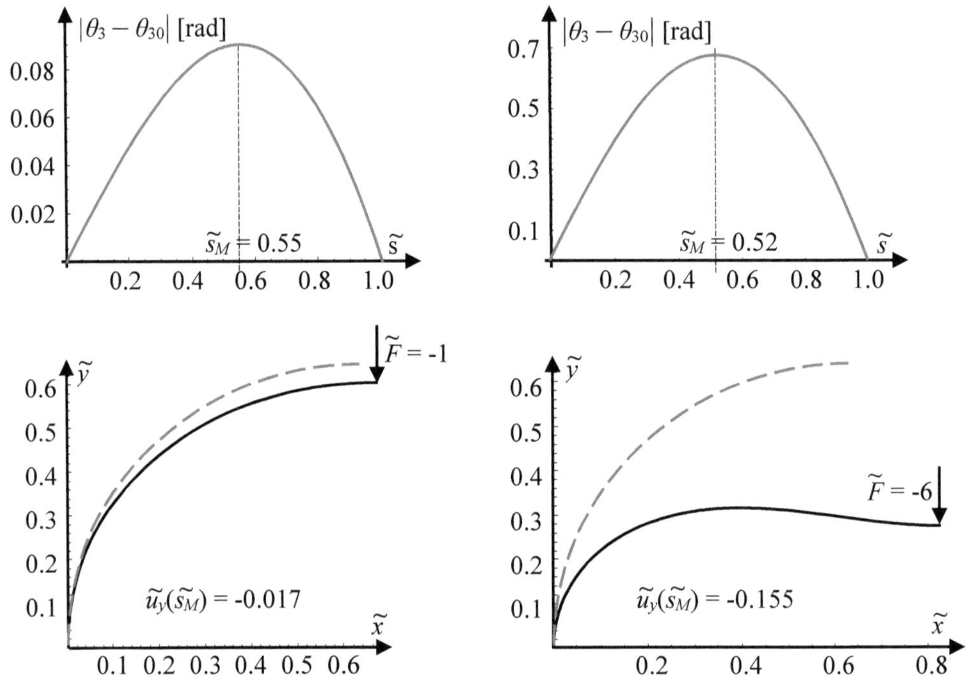

a b

Fig. 5.22: Difference $|\theta_3 - \theta_{30}|$ and the shapes of the ring segment before deformation (dotted line) and after deformation for the rod model from Fig. 5.20 e; a – for a dimensionless force with a value of -1; b – for a dimensionless force with a value of -6

Note that the force and the curvature are here both negative values. The following applies for the original shape of the rod:

$$\kappa_{30} = -\frac{1}{R}, \quad \theta_{30} = \frac{\pi}{2} - \frac{s}{R}. \tag{5.129}$$

The actuating force was increased until a maximum value of the angular difference $|\theta_3 - \theta_{30}|$ of around 0.7 rad (around 40°) was reached. Fig. 5.22 shows the solutions for the change in angle, as well as the shape of a ring segment of the gripping body both before and after a deformation due to a force in the Cartesian coordinate system. A suitable interval for the attachment of the fingers on the ring segment of the gripping body is as follows:

$$0.52 \le \tilde{s}_M \le 0.55\,. \tag{5.130}$$

Fig. 5.23 shows both variants of the acting force; a much smaller force is required for the second variant, in order to reach the maximum angular difference of around 40°. The area for the attachment of the fingers is also smaller. To find the *stroke* of a finger H, the length of the finger L_F, the displacement of the attachment point of the body $u_y(s_M)$ and the angle of the fingers α should be considered (Fig. 5.23). These parameters can be used here in dimensionless form.

$$\tilde{H} = \tilde{u}_y(\tilde{s}_M) + \tilde{L}_F\,|\theta_3 - \theta_{30}|\sin(\alpha - \frac{|\theta_3 - \theta_{30}|}{2}) \tag{5.131}$$

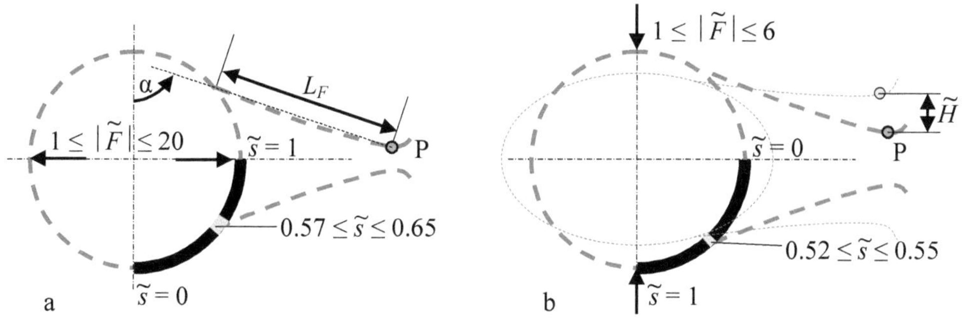

Fig. 5.23: Schematic depiction of the interval for the attachment of the fingers on the gripping body, and a possible shape of the finger for two variants of the actuating force

The displacement of the attachment point of the gripping device u_y is negative; greater values lead therefore to a smaller *stroke* of the fingers, according to (5.131). If the difference $|\theta_3 - \theta_{30}| \approx 0.7$ rad, as depicted in Fig. 5.21 b and Fig. 5.22 b, and equal finger lengths and their inclinations α in both cases are assumed, then the gripping device in Fig. 5.20 d or Fig. 5.23 b exhibits a greater stroke of the fingers. The same gripping device also has higher energy efficiency than the gripping device in Fig. 5.23 a due to the lower force. The gripping force is the same for both gripping devices and depends only from the size of the object to be gripped, because the object is held by the inherent elasticity of the gripping body and thus without energy input.

5.3.4 Two compliant mechanisms

Two compliant mechanisms with prismatic flexure hinges (compare with Table 2.2) are developed based on the model of a slider-crank mechanism (see Section 6.1) and each has three flexure hinges (Fig. 5.24). The first mechanism has rectangular hinge contours (compare with Table 6.2), while the second has semicircular hinges that are formed from semicircular cut-outs. The lengths and the minimum height of the hinges in both mechanisms are equal, the latter is denoted with h_0. The links of the mechanisms, as measurements of the sections between hinges, are equal in each mechanism and have a height of h_1. The lengths of the flexure hinges in both mechanisms are also equal and denoted with L_0. The values L_i denote the points on the mechanisms where either the curvature κ_{30} or the height $h(s)$ change. The behavior of each mechanism will be investigated and compared.

The regarded mechanism type is a partially compliant mechanism (compare with Fig. 2.8) that includes a slider element. The slider is actuated with a force **F**, which also causes a deformation of the compliant mechanism as a whole. This force constitutes both a direction-constant and follower load on the system, equations can therefore be formulated in either the attached or Cartesian coordinate system. Here, equations for the Cartesian coordinate system are taken from Section 4.7.6. The first two equations from (4.173) are solved analytically in conjunction with the boundary conditions.

$$Q_x(L_{10}) = -F, \quad Q_y(L_{10}) = 0 \tag{5.132}$$

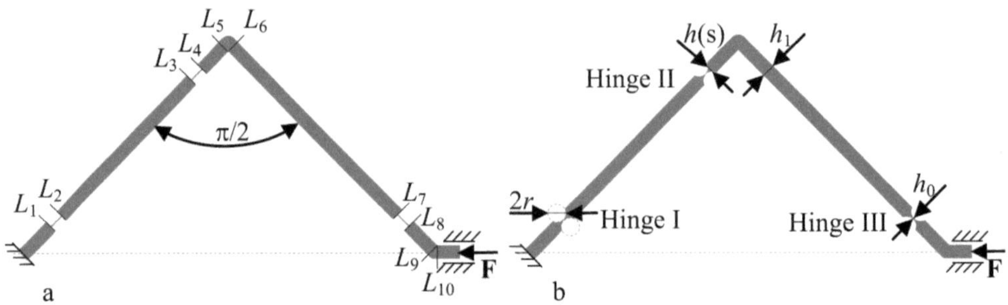

Fig. 5.24: Schematic of the mechanisms with geometric dimensions, not drawn to scale with (5.135); a – a mechanism with rectangular flexure hinges (M1); b – a mechanism with semicircular hinges (M2) formed from circular cut-outs

This result is then applied to the third equation in (4.173). The following equation system describes a deformed shape of the mechanism:

$$\begin{aligned}
&\frac{d\widetilde{M}_z}{d\tilde{s}} = \widetilde{F}\sin\theta_3,\\
&\frac{d\theta_3}{d\tilde{s}} = \frac{\widetilde{M}_z}{\widetilde{EI}_3} + \tilde{\kappa}_{30},\\
&\frac{d\tilde{x}}{d\tilde{s}} = \cos\theta_3,\\
&\frac{d\tilde{y}}{d\tilde{s}} = \sin\theta_3.
\end{aligned} \tag{5.133}$$

Dimensionless parameters are introduced to obtain universally valid conclusions from the results that are quantitatively dependent neither on the geometric or material parameters:

$$\begin{aligned}
&\tilde{L}_0 = \frac{L_0}{L_0} = 1, \quad \tilde{L}_i = \frac{L_i}{L_0}, i = 1,\dots,10, \quad \tilde{\kappa}_{30} = \kappa_{30}L_0,\\
&\tilde{s} = \frac{s}{L_0}, \quad \tilde{x} = \frac{x}{L_0}, \quad \tilde{y} = \frac{y}{L_0}, \quad \tilde{h}_1 = \frac{h_1}{L_0}, \quad \tilde{h}_0 = \frac{h_0}{L_0},\\
&\widetilde{F} = \frac{FL_0^2}{EI_3(0)}, \quad \widetilde{M}_z = \frac{FL_0}{EI_3(0)}, \quad \widetilde{EI}_3 = \frac{EI_3(s)}{EI_3(0)} = \frac{h^3(s)}{h_1^3}.
\end{aligned} \tag{5.134}$$

These parameters are selected for both mechanisms as follows:

$$\begin{aligned}
&\tilde{L}_i = \{1, 2, 8, 9, 10, 10.13, 18.09, 19.09, 20.09, 20.16\},\\
&\tilde{\kappa}_{30} = \begin{cases} 0 \quad \text{for} \quad 0 \le \tilde{s} \le \tilde{L}_5, \tilde{L}_6 \le \tilde{s} \le \tilde{L}_9 \\ -12.5 \quad \text{for} \quad \tilde{L}_5 \le \tilde{s} \le \tilde{L}_6 \\ 12.5 \quad \text{for} \quad \tilde{L}_9 \le \tilde{s} \le \tilde{L}_{10} \end{cases}\\
&\tilde{h}_0 = 0.1, \quad \tilde{h}_1 = 1.
\end{aligned} \tag{5.135}$$

A value of 0.08 is selected for the radii of curvature. The link lengths (left in Fig. 5.24) have the given values, and the remaining link lengths are calculated in such a way that the moving direction of the sliding element passes through the clamping point of the left end. The flexure hinges of these two mechanisms are shaped differently: the parameter h_0 describes the height of the hinge in the first mechanism (M1) with rectangular hinges; the parameter $h(s)$ describes the height of the hinge of the second mechanism (M2), with hinges formed with circular cut-outs with a radius r. The dimensionless values of these parameters are:

$$\tilde{r} = \frac{(\tilde{h}_1 - \tilde{h}_0)^2 + 1}{4(\tilde{h}_1 - \tilde{h}_0)},$$

$$\tilde{h}(s) = \begin{cases} \tilde{h}_0 + 2\tilde{r} - 2\sqrt{\left|\tilde{r}^2 - \left(\tilde{s} - \tilde{L}_1 - \frac{1}{2}\right)^2\right|}, & \text{for hinge I,} \\ \tilde{h}_0 + 2\tilde{r} - 2\sqrt{\left|\tilde{r}^2 - \left(\tilde{s} - \tilde{L}_3 - \frac{1}{2}\right)^2\right|}, & \text{for hinge II,} \\ \tilde{h}_0 + 2\tilde{r} - 2\sqrt{\left|\tilde{r}^2 - \left(\tilde{s} - \tilde{L}_8 - \frac{1}{2}\right)^2\right|}, & \text{for hinge III.} \end{cases} \tag{5.136}$$

The equations in (5.133) are solved with the following boundary conditions:

$$\begin{aligned} &\theta_3(0) = \frac{\pi}{4}, \quad \theta_3(L_{10}) = 0, \\ &\tilde{x}(0) = 0, \quad \tilde{y}(0) = \tilde{y}(L_{10}). \end{aligned} \tag{5.137}$$

The slider elements of the both mechanisms are now displaced by two dimensionless units (Fig. 5.25 a). Different forces are therefore required. These values for each mechanism, as well as the values of forces half as small, can be read and compared using Fig. 5.25 b.

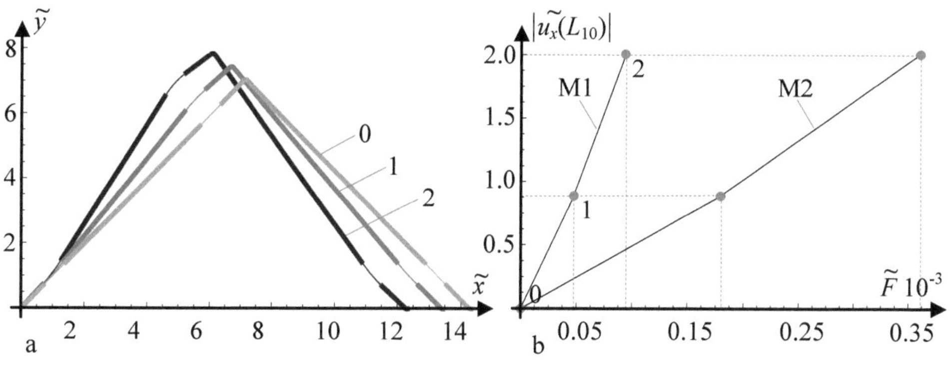

Fig. 5.25: Displacement of the mechanisms; a – unloaded state (0) and states (1 and 2) of a mechanism with rectangular flexure hinges after forces are applied; b – relationship between the deformation in the prismatic hinge and the actuating force for both mechanisms; M1 – mechanism with rectangular hinges, M2 – mechanism with semicircular hinges

The deformed states of both mechanisms are very similar and can barely be visually differentiated. In order to be able to compare these deflections more exactly, the bending angle θ_3 and the stress (for outer edge) from (5.138) are calculated and graphed (Fig. 5.26).

$$\tilde{\sigma}(s) = \frac{\widetilde{M}_z}{2\widetilde{EI}_3}\tilde{h}(s) \tag{5.138}$$

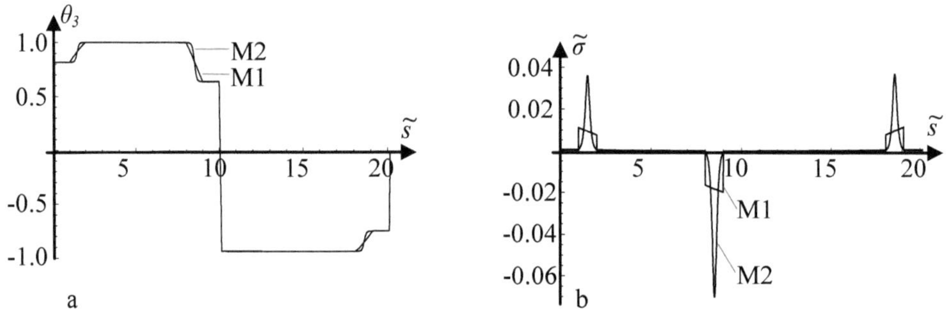

Fig. 5.26: Graph of the parameters for the mechanism with rectangular hinges (M1) and for the mechanism with semicircular hinges (M2); a – bending angle θ_3; b – stress distribution from (5.138)

For the same deflection, the mechanism with rectangular joints requires a force around three-and-a-half times smaller (Fig. 5.25 b) and thus bending stress more than three-and-a-half times smaller than in the second mechanism (Fig. 5.26 b). The deflection of the hinge regions is almost uniformly pronounced in the mechanism with rectangular joints, whereas the joints of the second mechanism undergo the greatest deflection in the middle (Fig. 5.26 a). This fact is confirmed by the stress distribution within the joint region in Fig. 5.26 b.

5.4 Spatial problems in the Cartesian coordinate system

The equations listed in Section 4.7.5 can be used for spatial problems with direction-constant loads. The equations for the forces (4.162) can mostly be integrated analytically, thus reducing the number of differential equations that must be solved numerically. Especially where there is no distributed force and the first three boundary conditions from (4.168) are used, the solutions for the forces can be expressed as simple constants in most cases. In the following examples, a valve with compliant elements and a gripping device with compliant fingers under the effect of direction-constant loads are modeled and investigated.

5.4.1 A compliant valve

A compliant valve consists of three curved compliant connections and a cylindrical pin, which moves under external force and closes an opening (Fig. 5.27 a). The connections join the pin with the bottom. The pin moves only in the axial direction by mechanical guidance. The value of the pin radius can be neglected. Under assumption, that three connections exhibit the similar movement, only one connection can be considered. The compliant connection has the form of a planar *helix* with a length L and is clamped at one end. The initial form, including the curvature κ_{30}, of the helix is given as follows:

$$\kappa_{10} = 0, \quad \kappa_{20} = 0,$$
$$\kappa_{30} = \frac{1}{100}\left(1 + \frac{s}{10}\right). \tag{5.139}$$

In general, the parameters κ_{i0} and $\kappa_{(i0)}$ have the same value for rods with rotationally symmetric cross-sections. The force of the pin is $3\mathbf{F_z}$. Therefore, the helix is loaded with an external force $\mathbf{F_z}$ on its end, which is jointed with the pin (Fig. 5.27 a). Due to force acting, the connection takes on a three-dimensional form. The behavior of the helical valve connection will be described.

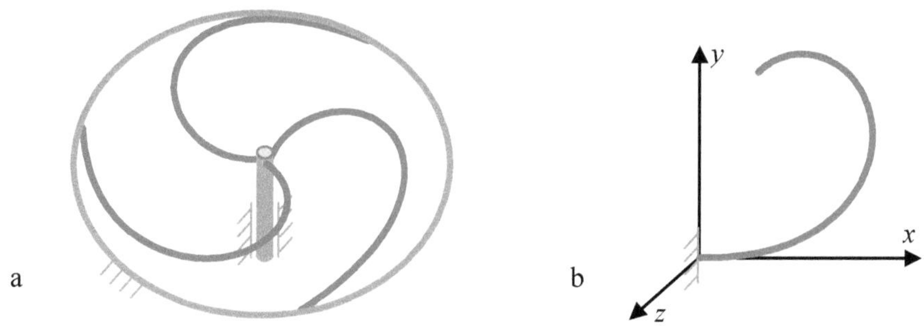

Fig. 5.27: A compliant valve with three curved compliant connections; a – the valve; b – one connection of the valve

The deformed state of the rod is modeled using the equations from Section 4.7.5, which have been adapted for use here:

$$\frac{dQ_x}{ds} = 0,$$
$$\frac{dQ_y}{ds} = 0, \tag{5.140}$$
$$\frac{dQ_z}{ds} = 0.$$

Not all equations from (5.140) can be integrated analytically, because the number of boundary conditions is insufficient.

$$Q_z(L) = F_z \tag{5.141}$$

Then the equations for the moments from (4.163) are needed:

$$\frac{dM_x}{ds} + Q_z t_{12} - Q_y t_{13} = 0,$$
$$\frac{dM_y}{ds} - Q_z t_{11} + Q_x t_{13} = 0, \tag{5.142}$$
$$\frac{dM_z}{ds} + Q_y t_{11} - Q_x t_{12} = 0.$$

The equations from (4.164) and (4.170) are used alongside the equations in (5.142) to describe the shape of the rod. The following boundary conditions are used:

$$\begin{array}{llll} M_z(L)=0, & \theta_1(0)=0, & x(0)=0, & x(L)=x_0(L), \\ \theta_1(L)=0, & \theta_2(0)=0, & y(0)=0, & \\ \theta_2(L)=0, & \theta_3(0)=0, & z(0)=0. & y(L)=y_0(L). \end{array} \tag{5.143}$$

Thus, the twelve boundary conditions (5.143) and (5.141) are available to solve the twelve differential equations (5.140), (5.142), (4.164) and (4.170). The parameter $x_0(L)$ and $y_0(L)$ are coordinates x and y at the end of the helix for its initial shape. They can be obtained by the parameter relationships from (4.104) (or (4.164)) and (4.170) used for the initial shape of the helix:

$$\begin{aligned} \kappa_{10} &= \frac{d\theta_{10}}{ds} - \sin\theta_{20}\frac{d\theta_{30}}{ds}, \\ \kappa_{20} &= \cos\theta_{10}\frac{d\theta_{20}}{ds} + \sin\theta_{10}\cos\theta_{20}\frac{d\theta_{30}}{ds}, \\ \kappa_{30} &= -\sin\theta_{10}\frac{d\theta_{20}}{ds} + \cos\theta_{10}\cos\theta_{20}\frac{d\theta_{30}}{ds}, \\ &\frac{dx_0}{ds} - \cos\theta_{20}\cos\theta_{30} = 0, \\ &\frac{dy_0}{ds} - \cos\theta_{20}\sin\theta_{30} = 0, \\ &\frac{dz_0}{ds} + \sin\theta_{20} = 0 \end{aligned} \tag{5.144}$$

with boundary conditions for the clamped end:

$$\begin{array}{ll} \theta_{10}(0)=0, & x_0(0)=0, \\ \theta_{20}(0)=0, & y_0(0)=0, \\ \theta_{30}(0)=0, & z_0(0)=0. \end{array} \tag{5.145}$$

The following expressions are used for the parameters κ_i, according to the equations (4.166):

$$\begin{aligned} \kappa_1 &= \frac{1}{GI_1}\left(M_x t_{11} + M_y t_{12} + M_z t_{13}\right), \\ \kappa_2 &= \frac{1}{EI_2}\left(M_x t_{21} + M_y t_{22} + M_z t_{23}\right), \\ \kappa_3 &= \frac{1}{EI_3}\left(M_x t_{31} + M_y t_{32} + M_z t_{33}\right) + \kappa_{(30)}. \end{aligned} \tag{5.146}$$

The following parameters of the elastic steel rod are used, which has a length L and a circular cross-section with a radius r:

$$r = 0.5\ \text{mm}, \quad L = 80\ \text{mm},$$
$$F_{z1} = 1\ \text{N}, \quad F_{z2} = 2\ \text{N}, \quad F_{z3} = 3\ \text{N}, \tag{5.147}$$
$$E = 2.1 \cdot 10^5\ \frac{\text{N}}{\text{mm}^3}, \quad G = 7.93 \cdot 10^4\ \frac{\text{N}}{\text{mm}^3}.$$

The helix rod is acted by three forces F_{z1}, F_{z2} and F_{z3}.

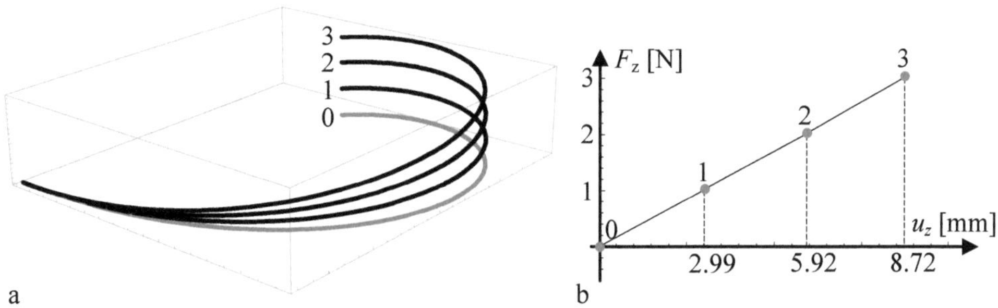

Fig. 5.28: Deformation of valve connection; a – initial shape (0) and three deformed shapes (1, 2 and 3) under the effect of different forces; b – force-displacement characteristic

The initial shape and three deformed shapes are shown in Fig. 5.28 a. The force-displacement characteristic (Fig. 5.28 b) is approximately linear.

5.4.2 A gripping tool with curved compliant fingers

A compliant element in the form of a quarter segment of the ring with a radius of curvature R and a circular cross-section with radius r is used as a gripping finger. The aim here is to investigate how the positioning of such fingers in a gripping device has an effect on their deformations. Three different positions of the gripping fingers to be investigated are given in Fig. 5.29 a-c.

The gripping fingers move towards one another until a certain gripping force is reached, meaning the gripping normal force in each of the three cases is the same. A gripping finger is modeled as a compliant element clamped at one end in the form of a ring segment under the effect of a gripping force. This gripping force consists of a normal force $\mathbf{F_N}$ and a friction force $\mathbf{F_R}$. The direction of this normal force is different in each case for the model of the ring segment, but remains constant in value (Fig. 5.30). The friction force exerted by the object on the finger has the same direction as the gravity $\mathbf{g}$ and is limited by a coefficient μ_0 for static friction (5.148).

$$F_R \le \mu_0 F_N \tag{5.148}$$

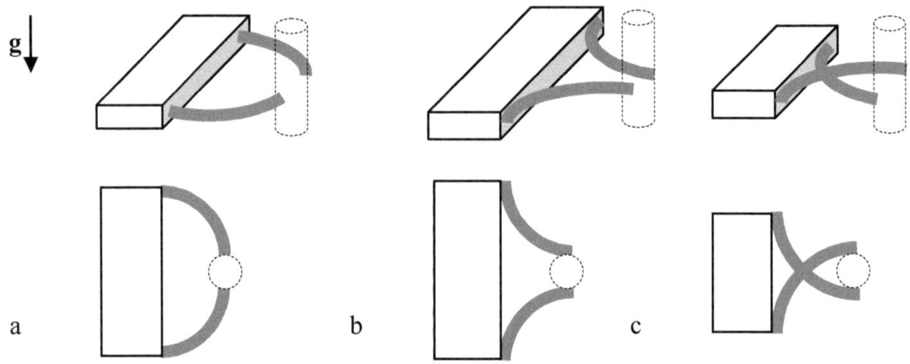

Fig. 5.29: Three different positions of the gripping fingers a-c; above – three-dimensional view of the gripping device; below – a view of the plane orthogonal to the acceleration of gravity **g**

Here only the limit value for F_R is considered. For this investigation, three models of the gripping fingers are calculated and the deformations of the fingers are compared to one another. Equations from Section 4.7.5 are used for this purpose. First, the equations from (4.162) are solved analytically for all three cases, resulting in the internal forces:

$$
\begin{aligned}
&\text{Case a: } Q_x = 0, \quad Q_y = -F_N, \quad Q_z = \mu_0 F_N, \\
&\text{Case b: } Q_x = -F_N, \quad Q_y = 0, \quad Q_z = \mu_0 F_N, \\
&\text{Case c: } Q_x = F_N, \quad Q_y = 0, \quad Q_z = \mu_0 F_N.
\end{aligned} \tag{5.149}
$$

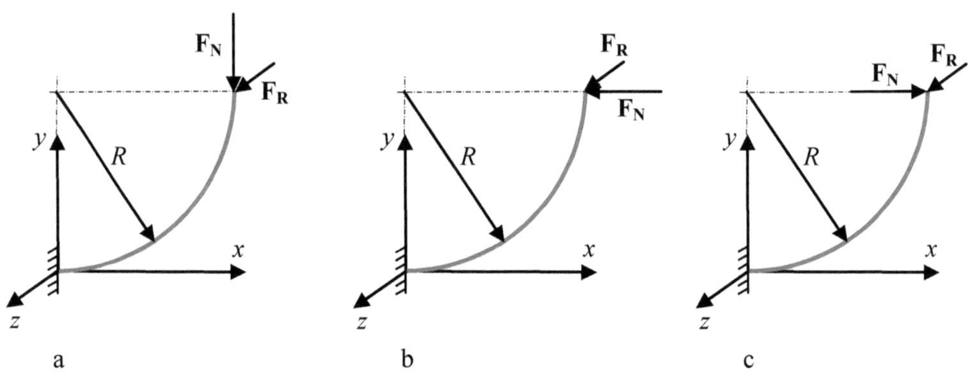

Fig. 5.30: Three models for three cases a-c of a gripping finger as a ring segment clamped at one end as in Fig. 5.29 a-c; $\mathbf{F_R}$ – a friction force, $\mathbf{F_N}$ – a normal force

Dimensionless parameters are introduced, in order to provide universally applicable results. The chosen examples for the introduction of these dimensionless parameters for the lengths, forces, moments and stiffnesses are given below:

$$\tilde{L} = \frac{L}{L} = 1, \quad \tilde{s} = \frac{s}{L}, \quad \tilde{F} = \frac{FL^2}{EI_3}, \quad \tilde{M} = \frac{ML}{EI_3},$$
$$\widetilde{EI}_3 = \frac{EI_3}{EI_3} = 1, \qquad \widetilde{EI}_2 = \frac{EI_2}{EI_3} = 1, \qquad \widetilde{GI}_1 = \frac{GI_1}{EI_3}. \tag{5.150}$$

The solutions for the internal forces from (5.149) are applied to the equations from (4.163). For the first case in Fig. 5.30 a, the equations for the moments with dimensionless parameters assume the following form:

$$\frac{d\tilde{M}_x}{d\tilde{s}} + \mu_0 \tilde{F}_N t_{12} + \tilde{F}_N t_{13} = 0,$$
$$\frac{d\tilde{M}_y}{d\tilde{s}} - \mu_0 \tilde{F}_N t_{11} = 0, \tag{5.151}$$
$$\frac{d\tilde{M}_z}{d\tilde{s}} - \tilde{F}_N t_{11} = 0.$$

The following applies for the moments from (5.151), according to (4.167):

$$\tilde{M}_x = t_{11}\widetilde{GI}_1\tilde{\kappa}_1 + t_{21}\tilde{\kappa}_2 + t_{31}(\tilde{\kappa}_3 - \tilde{\kappa}_{(30)}),$$
$$\tilde{M}_y = t_{12}\widetilde{GI}_1\tilde{\kappa}_1 + t_{22}\tilde{\kappa}_2 + t_{32}(\tilde{\kappa}_3 - \tilde{\kappa}_{(30)}), \tag{5.152}$$
$$\tilde{M}_z = t_{13}\widetilde{GI}_1\tilde{\kappa}_1 + t_{23}\tilde{\kappa}_2 + t_{33}(\tilde{\kappa}_3 - \tilde{\kappa}_{(30)}).$$

The following parameters are applied to describe the unloaded shape of the gripping finger:

$$\tilde{\kappa}_{(10)} = 0,$$
$$\tilde{\kappa}_{(20)} = 0, \tag{5.153}$$
$$\tilde{\kappa}_{(30)} = \frac{1}{\tilde{R}}.$$

Six further differential equations based on the equations in (4.164) and (4.170) can be used to determine the shape of the finger in the Cartesian coordinate system for dimensionless parameters:

$$\frac{d\theta_1}{d\tilde{s}} = \tilde{\kappa}_1 + \tilde{\kappa}_2 \sin\theta_1 \tan\theta_2 + \tilde{\kappa}_3 \cos\theta_1 \tan\theta_2,$$
$$\frac{d\theta_2}{d\tilde{s}} = \tilde{\kappa}_2 \cos\theta_1 - \tilde{\kappa}_3 \sin\theta_1, \tag{5.154}$$
$$\frac{d\theta_3}{d\tilde{s}} = \tilde{\kappa}_2 \sin\theta_1 \sec\theta_2 + \tilde{\kappa}_3 \cos\theta_1 \sec\theta_2,$$

$$\frac{d\tilde{x}}{d\tilde{s}} = \cos\theta_2 \cos\theta_3,$$
$$\frac{d\tilde{y}}{d\tilde{s}} = \cos\theta_2 \sin\theta_3, \qquad (5.155)$$
$$\frac{d\tilde{z}}{d\tilde{s}} = -\sin\theta_2.$$

Nine boundary conditions can be used to solve the nine differential equations (5.151) with the moments from (5.152), (5.154) and (5.155):

$$\begin{aligned} &\tilde{\kappa}_1(L) = 0, && \theta_1(0) = 0, && \tilde{x}(0) = 0, \\ &\tilde{\kappa}_2(L) = 0, && \theta_2(0) = 0, && \tilde{y}(0) = 0, \\ &\tilde{\kappa}_3(L) = \frac{1}{\tilde{R}}, && \theta_3(0) = 0, && \tilde{z}(0) = 0. \end{aligned} \qquad (5.156)$$

Three dimensionless parameters used to solve this problem are written as follows for gripping fingers made of steel:

$$\begin{aligned} &\tilde{L} = 1, \quad \tilde{r} = 0.1, \quad \tilde{R} = \frac{2}{\pi}, \quad \mu_0 = 0.25, \quad \tilde{F}_N = 1, \\ &\tilde{F}_R = \mu_0 \tilde{F}_N, \quad \widetilde{EI}_3 = 1, \quad \widetilde{EI}_2 = 1, \quad \widetilde{GI}_1 = 0.755. \end{aligned} \qquad (5.157)$$

The results of solving the differential equations with corresponding boundary conditions from (5.156) for the three cases in (Fig. 5.30) are given in Fig. 5.31 and summarized in Table 5.1. This shows a complete displacement of the end of the finger and its displacement in the z-direction.

The complete displacement of the finger can be calculated by using the resulting coordinates of the end of the finger:

$$\tilde{u} = \sqrt{(\tilde{x}(1) - \tilde{R})^2 + (\tilde{y}(1) - \tilde{R})^2 + (\tilde{z}(1))^2}\,. \qquad (5.158)$$

The following applies for the displacement in the z-direction:

$$\tilde{u}_z = \tilde{z}(1)\,. \qquad (5.159)$$

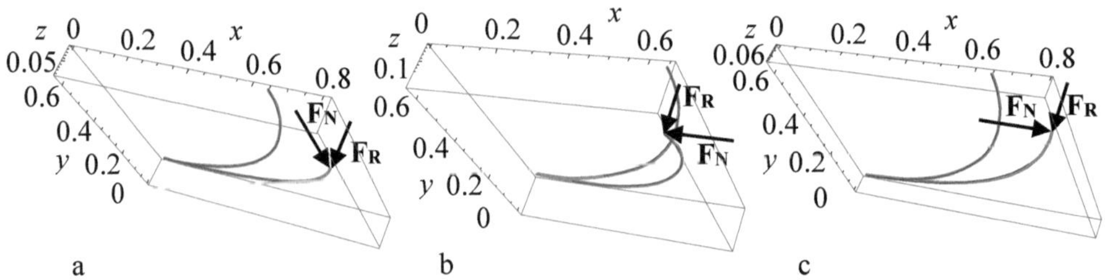

Fig. 5.31: Unloaded shape and deformed shape of the gripping fingers in Fig. 5.30 for three different load cases under the effects of the normal force and friction force

Table 5.1: Results for the displacements of a curved gripping finger as dimensionless values; a complete displacement in space and a displacement in the z-direction for the three load cases in Fig. 5.30 are shown; the last column includes the values for these displacements under the effect of the force $\mathbf{F_R}$ ($\mathbf{F_N} = \mathbf{0}$)

No	Calculated values	Case a	Case b	Case c	Under acting of force $\mathbf{F_R}$ ($\mathbf{F_N} = \mathbf{0}$)
1	Complete displacement	0.2559	0.2755	0.2082	0.0806
2	Displacement in the direction of z	0.1185	0.0900	0.0641	0.0805

In all three cases, the absolute value of the complete displacement of the gripping fingers is different while the gripping force in each of the three cases is the same. The values for the displacement in the z-direction are also different from one another. These values are not dependent on the value of the complete displacement, meaning that the greatest complete displacement (case b) does not correspond with the greatest displacement in the z-direction (case a). However, case c exhibits the smallest displacements, both in the complete displacement and the displacement in the z-direction.

If the compliance is considered as a quotient of the complete displacement and the force, the finger in case b has the greatest compliance, whereas the finger in case c has the least. In order to compare the results for all three cases with the displacement of the fingers under the effects of the friction force alone, the last column of Table 5.1 give the values for this load ($F_R \neq 0$, $F_N = 0$) for all cases. The displacement in the z-direction under the effect of the force $\mathbf{F_R}$ (last column of Table 5.1) is greater than in case c and smaller than in cases a and b. The complete displacement of the end of the finger under the effect of the force $\mathbf{F_R}$ alone is, as expected, smaller than the complete displacement in cases a-c.

If the smallest possible complete displacement of the gripping fingers or their displacements in the z-direction is desired, positioning the gripping fingers as in case c from Fig. 5.29 is recommended. If, however, a higher compliance of the fingers in relation to the complete displacement is desired, case b should be selected.

6 Synthesis of compliant mechanisms and design of flexure hinges

In the previous chapters, mostly compliant systems and structures with distributed compliance were in the focus of investigation. In high-precision motion systems, however, often compliant mechanisms with concentrated compliance are being used, too. In these compliant mechanisms [49], mostly notch flexure hinges [68] are used as materially coherent revolute joints due to their smooth and repeatable motion. But because of the elastic bending the rotation is approximated and limited. In literature, numerous different cut-out geometries have been used to describe the contour of notch hinges, while circular or corner-filleted contours are mainly used in technical systems. Many more types of flexure hinges are possible, which will be presented in this chapter. However, accurately predicting the contour-dependent deformation behavior, stress behavior, and especially the motion behavior, of a flexure hinge is challenging without carrying out FEM simulations.

In this chapter, an angle-based synthesis method for compliant mechanisms with notch flexure hinges based on the rigid-body replacement approach is suggested, and novel design guidelines for individually shaped hinge contours within one mechanism are presented for commonly-used and polynomial flexure hinges. Polynomial flexure hinges are particularly suitable here, because they allow for a simple CAD-based modeling and, due to an adjustable order, a variable and goal-oriented design of the compliant mechanism.

Using the example of a fully compliant four-bar mechanism, it will be demonstrated that this synthesis method can be used to design highly precise compliant mechanisms with a large stroke. This synthesis method is applicable for notch hinges in both fully and partially compliant mechanisms. The design graphs and design equations for this method can accelerate the design of compliant mechanisms with optimized notch flexure hinges.

6.1 Rigid-body replacement approach

Three main approaches exist for the *synthesis* of a compliant mechanism:

- synthesis through the rigid-body-replacement method (e.g. [48]);
- synthesis through the topology optimization method (e.g. [31]);
- synthesis through constrained-based methods, such as the freedom and constraint topologies method (e.g. [45]) or the building blocks method (e.g. [2]).

Depending on the approach, compliant mechanisms with different structural designs and compliance distributions are induced. The *rigid-body replacement method* is more suitable than the topology optimization synthesis, since the former method provides better point guidance accuracy [84]. Starting with the *rigid-body mechanism*, this leads to an iterative design process for a *compliant mechanism* exhibiting mostly concentrated compliance based on notch flexure hinges.

https://doi.org/10.1515/9783110479744-143

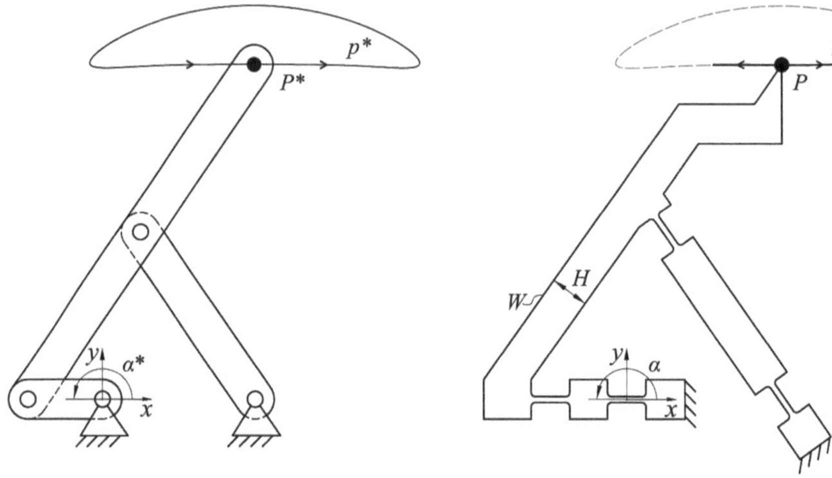

Fig. 6.1: Synthesis of a compliant mechanism using the rigid-body replacement method; a – rigid-body mechanism; b – compliant mechanism with corner-filleted notch flexure hinges (with height H and width W)

Regarding the typical example of a four-bar mechanism after HOECKEN realizing an approximated straight-line path p of a coupler point P [47], the rigid-body mechanism and compliant counterpart are shown in Fig. 6.1 in replacement position $\alpha = \alpha^*$.

The sequential procedure of type synthesis, dimensional synthesis and constructional design, which is often used for rigid-body mechanisms, cannot be applied to compliant mechanisms, due to the problem of continuum mechanics. The principal reason here being that the compliant mechanism must exhibit special force-displacement characteristics, while still carrying out the required motion task. Thus, both kinematic and kinetic behaviors must be considered simultaneously as multi-objective design criteria, while still taking structural strength into account. Based on these requirements, synthesis using rigid-body replacement method can be divided into four basic steps:

i. synthesis of a suitable rigid-body mechanism;
ii. replacement of the hinges and basic design of the compliant mechanism;
iii. goal-oriented, specific geometric design of the flexure hinges and
iv. verification of results and proof of requirements.

Regarding the required mechanism properties, the geometric hinge design is a key step of the synthesis of a compliant mechanism using the rigid-body replacement method. In previous literature, the specific geometric design of the flexure hinges during synthesis is only considered when using standard contours [76] with almost identical hinges in one compliant mechanism. Thus, in addition to existing approaches, such as using complex flexure hinge types within the mechanism (e.g. [27], [110]) or increasing the hinge number of the kinematic chain (e.g. [3], [22]), optimizing the contours of incorporated notch flexure hinges may prove useful in the synthesis of compliant mechanisms. This rigid-body replacement approach can be applied to planar and spatial mechanisms, but the following explanations are focused on planar compliant mechanisms with revolute joints.

The differentiation between concentrated and distributed compliance is usually based on experience, although a classification approach based on the maximum dimensions of the system or the mechanism has been previously described in Section 2.1. In addition, geometric relations are introduced, such as the ratio of the hinge length to the minimum notch height [20]. In this case, this definition is limited to a separate hinge without considering the lengths and cross-sections of adjacent links. It is therefore necessary to consider the size relationship between the flexure hinge and the compliant mechanism synthesized through the rigid-body replacement approach. As a result, the definition from Section 2.1 can be extended to include a comparison of the flexure hinge lengths to the characteristic link lengths of the rigid-body model (q.v. [58]).

6.2 Types of flexure hinges

A *flexure hinge* is seen as a compliant joint which approximately acts as a *hinge* due to flexural bending. Thus, the form of relative motion can be idealized as a rotation (cf. Table 2.3). Unlike form and force-closed joints, a flexure hinge provides a restoring force (a property called *bending stiffness* due to rotation), which can be advantageous in technical systems. Depending on the coherent connection, the angular deflection of a flexure hinge is limited by maximum acceptable material stresses or elastic strain values (*maximum angular deflection*). Thus, the *motion range* of a compliant mechanism is also limited by the hinge with the largest rotation angle in the kinematic chain, assuming the same contours are used. In addition, no exact relative rotation is possible with a flexure hinge, because its axis of rotation is always shifted as geometric and load parameters vary (*rotational precision*). In turn, this can lead to path deviations in the compliant mechanism compared with the *rigid-body model* that can no longer be considered negligible, especially in precision applications ([64], [102]).

In the following, as a flexure hinge a monolithic, small-length and elastic deformable segment of a compliant mechanism is defined, that provides a relative rotation of two adjacent links mainly limited due to bending stress. Because of their monolithic arrangement, compliant joints provide numerous approaches for the design of a flexure hinge when compared to a conventional form or force-closed revolute pair. Based on the well-described leaf-type flexure hinge, many different flexure hinge types have been developed in past decades or introduced in recent works in order to realize a larger angular deflection and/or a more precise rotation. Since the leaf-type hinge, which is formed from a simple strip, many elementary or complex flexure hinges have since been described in other academic works.

Based on literature, a selection of twenty-four *flexure hinge types* used to achieve a rotation with one desired degree of freedom is provided in Fig. 6.2: The leaf-type hinge [106], the curved leaf-type hinge [32], the folded leaf-type hinge [29], the spiral leaf-type hinge [100], the crossed (leaf-type) hinge [52], the curved crossed hinge [73], the stiffened crossed hinge [12], the prismatic crossed hinge or "cartwheel" hinge [5], the serially combined prismatic crossed hinge with self-compensation [75], the trapezoidal hinge [86], the multi-trapezoidal hinge or "butterfly" hinge [43] and the multi-leaf hinge with a centric axis [6]. For some hinges, especially the latter ones, the structural differentiation between a hinge and a mechanism is no longer obvious. Other types include the elastomer hinge [85] and the typical used notch hinge [83] with some similar hinges, such as the curved notch hinge [69], the leaf-type notch hinge [111], the multi-notch hinge with several inside notches [67] and the multi-notch hinge with several outside notches [72].

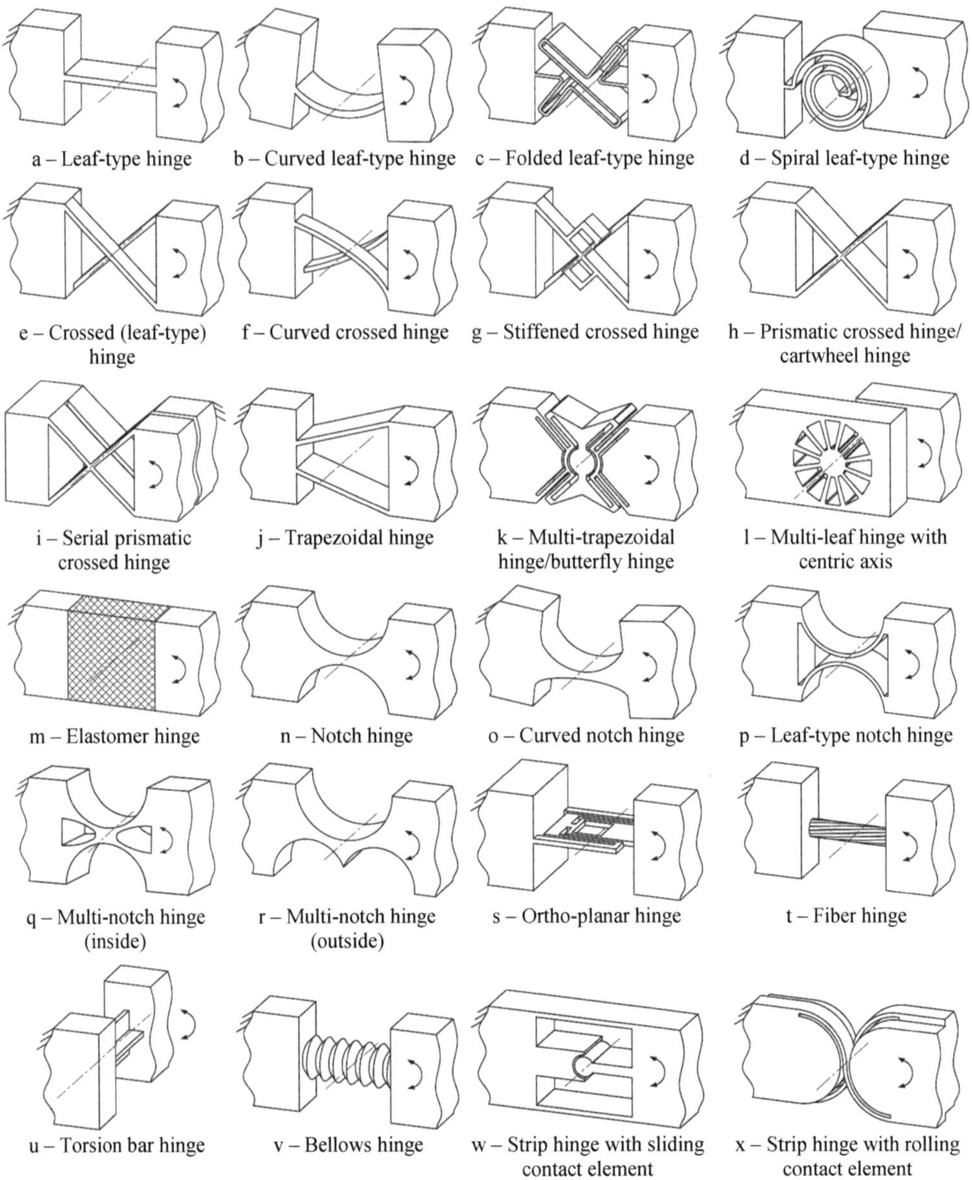

Fig. 6.2: Types of flexure hinges used to achieve rotational motion with a desired degree of freedom of $f = 1$

Hinges with a special structural characteristic are the ortho-planar hinge or “lamina emergent” hinge [104], the fiber hinge [82], the torsion bar hinge [97], the bellows hinge [91], the strip hinge with an additional sliding contact element [11] and the strip hinge with a rolling contact element [13]. Further not depicted flexure hinges with specific properties are used in compliant mechanisms, like the torsion spring hinge [81], the active hinge [109] or the lattice hinge [79].

Table 6.1: FEM-based characterization of ten relevant flexure hinges with comparable dimensions and the same leaf height ("++": very high/large; "+": high/large; "–": low/small; "––": very low/small; "x": true)

Hinge type	Hinge design			Hinge properties		
	Constant leaf height	Prismatic manufactur.	Complexity	Bending compliance	Angular stroke	Rotational precision
Leaf-type hinge 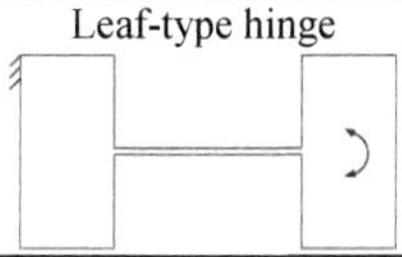	x	(x)	–	++	+	–
Corner-filleted notch hinge 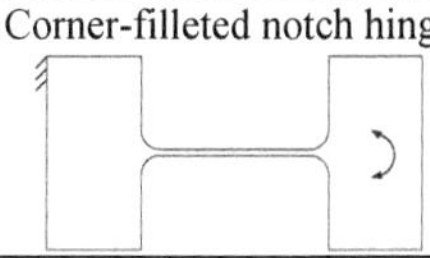		x	––	++	+	–
Semicircular notch hinge 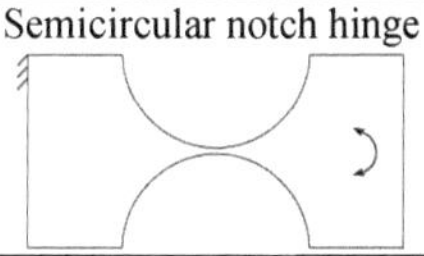		x	––	–	––	+
Leaf-type circular notch hinge 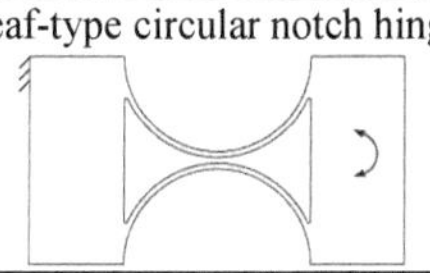	x	x	–	–	–	––
Symmetric crossed hinge 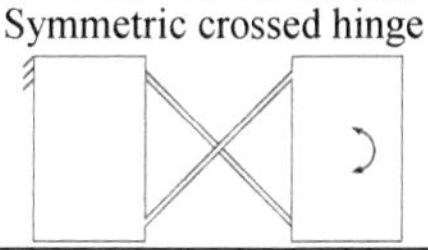	x		+	++	++	–
Asymmetric crossed hinge 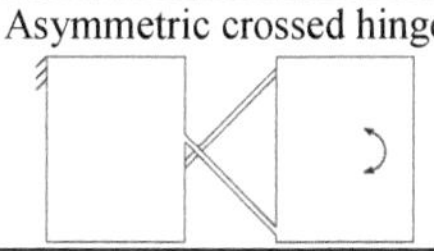	x		+	+	–	––
Prismatic crossed hinge 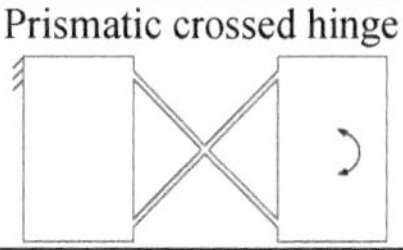	x	x	–	–	–	+
Serial prismatic crossed hinge 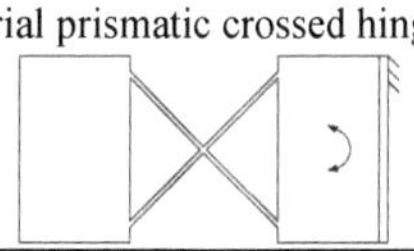	x		++	+	+	++
Trapezoidal hinge 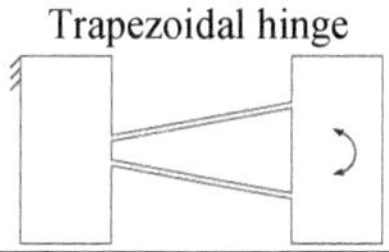	x	x	–	––	––	––
Multi-trapezoidal hinge 		x	++	+	+	++

Based on an FEM analysis [24], the elasto-kinematic hinge properties of ten different flexure hinges relevant in precision engineering applications are investigated. The results of a four-stage qualitative characterization of their bending compliance (as the reciprocal of bending stiffness), maximum angular deflection *stroke* (for reaching the *admissible strain* limit), and rotational precision (a model-based approach of guiding the center with a constant distance of the half hinge length is used, see [58]) are shown in Table 6.1.

Considering all three hinge properties, the results show that some hinges, such as the serial prismatic crossed hinge or the butterfly hinge, are especially suitable, although their manufacturing is complicated by their high complexity. Other hinges, such as the trapezoidal hinge, are not suitable for the parameter configuration that is the focus of this investigation. When comparing the corner-filleted and semicircular notch hinges to the remaining types, it becomes clear that different design goals can be met by selecting between comparable, simple notch hinge designs already, largely due to a great contour variety.

Thus, the following design guidelines in this chapter are focused on notch flexure hinges. Due to their low complexity, they are easy to manufacture and therefore mainly used in planar compliant mechanisms, especially in kinematic chains with a higher link number. Furthermore, with regard to the rotational precision and possible deflection stroke as contrary objectives, notch flexure hinges offer great potential for contour optimization, a method not yet employed. Hence, general design guidelines for common hinge contours or special high-performance hinge contours, such as polynomial flexure hinges, would be of great benefit.

6.3 Types of notch flexure hinges and suggested contour shapes

The rotation angle of *notch flexure hinges* is generally limited by the materially coherent design. In dependence of the material, angles over 70° can be realized with plastic film joints [85]. But for high-strength metals, which are typical for monolithic micro and nano-positioning systems, the rotation is limited to small angles of a few degrees [107]. Superelastic hinges enable angles of up to 30° for optimized *notch contours* [26]. In addition, notch flexure hinges have often been geometrically designed so that various cut-out geometries are proposed in literature to describe the variable contour height $h_c(x)$, see Fig. 6.3. A classification of notch hinges by their different contour shape characteristics is shown in Table 6.2.

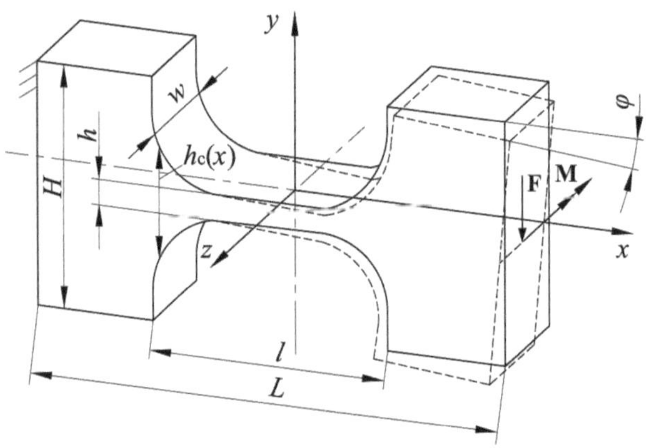

Fig. 6.3: Notch flexure hinge with its geometric parameters and the deflected state with φ as result of the load

Table 6.2: Classification of notch flexure hinges by their different contour shape characteristics

Criterion	Category	Subcategory	Example
Contour form/ notch geometry	Elementary shapes	Rectangular	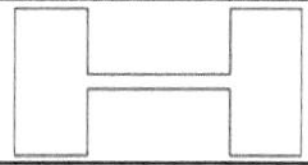
		Corner-filleted	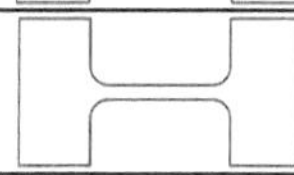
		Circular	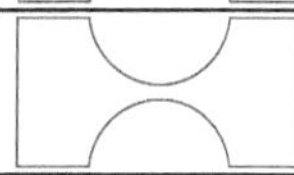
		Elliptical	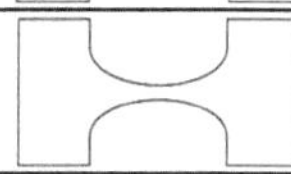
		Parabolic	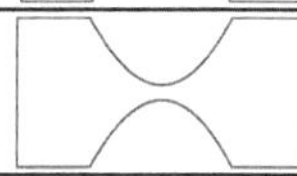
		Hyperbolic	
		Cycloidal	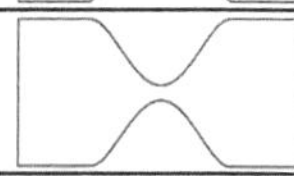
		Several combinations	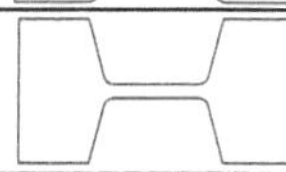
	Specific mathematical functions	Polynomial function of higher order	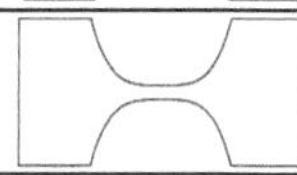
		Spline function	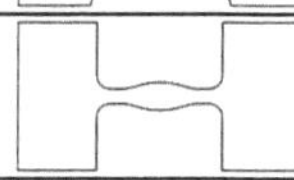
		Other (BEZIÉR function, trigonometric function, exponential function, …)	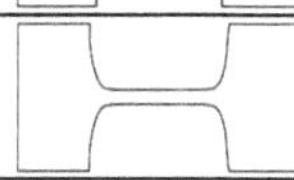
Contour intersection	Continuous		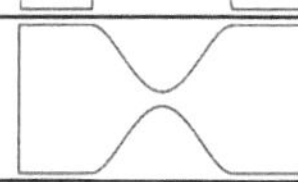
	Not continuous		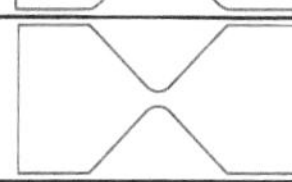
Contour symmetry	Symmetric		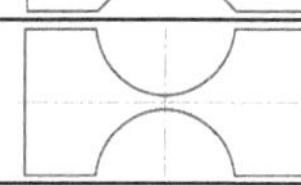
	Transversely symmetric		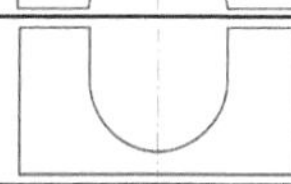
	Axially symmetric		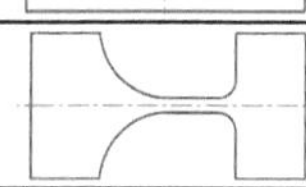
	Asymmetric		

Mostly, there are predefined basic geometry elements that lead to three main *notch hinge types*, each with a typical property: The precise hinge with a semicircular contour [83], the large-deflective hinge with a corner-filleted contour [71] or the elliptical hinge [93] as a compromise. Furthermore, flexure hinges are designed with other elementary geometries to realize a special property, like the parabolic contour [25], hyperbolic contour [70] and the cycloidal contour [96]. Increasingly, flexure hinges are being designed with a combination of these basic geometries [113]. Rarely, special mathematical functions are used that allow for more precise shape variations of the partial or whole hinge contour, due to a higher number of parameters, such as the spline contour [10], the power function contour [56], the exponent-sine contour [103], the LAMÉ contour [23] and the BÉZIER contour [99]. Recently, special mixed forms, such as the "z-shaped" flexure hinge [55], have also been investigated. The design with undefined freeform geometries based on topology optimization, as suggested in [124], requires a complex and unintuitive design process, which also makes it difficult to generalize.

Higher-order polynomial functions, as suggested by the author in [60], are rarely discussed in this context. Among the variety of cut-out geometries, however, polynomial functions in particular hold promise for contour optimization, while still allowing the contours to be modeled fairly simply. Depending on the polynomial order and the coefficients, arbitrary complex curves can be realized or nearly any elementary geometry can be approximated.

Although completely symmetric flexure hinges are usually used, there have been several studies of transversally and axially symmetric hinges. Axially symmetric hinges in particular are mostly realized as *hybrid flexure hinges*, because they combine the advantages of right circular and corner-filleted flexure hinges [17]. Furthermore, it has been shown that smaller radii at the loaded hinge side lead to improved kinematic behavior, until an ideal rotation axis is reached ([57], [59]). However, only fully symmetric notch hinges are considered for mechanism synthesis first, because they allow for both a holistic and intuitive design process.

Based on design of experiments of separate flexure hinges, it has been proven that the objectives depend on the basic hinge dimensions as detailed below ([87], [113]):

- the bending stiffness and maximum stress increase in particular as the minimum hinge height or notch height h increases,
- the rotational precision decreases with an increasing minimum notch height h.

With regard to general guidelines for the design of a suitable corner-filleted flexure hinge, discrete values for specific hinge dimensions providing low stresses are suggested:

- a ratio of the fillet radius r and the minimum notch height h with $r = 0.64h$ [105], $r = 0.7h$ [92], $r = h$ [95] or $r = 2h$ [39];
- a ratio of the fillet radius r and the hinge length l with $r = 0.1l$ ([59], [76], [92]).

For the design of a suitable circular flexure hinge, the following ratios are recommended:

- a ratio of the circular radius R and the minimum notch height h with $R > 5h$ [42];
- a ratio of the minimum notch height h and the link height H with $h < 0.3H$ [88].

Sometimes, special contours are suggested as a result of optimizing the contours of a separate flexure hinge or a hinge applied in the mechanism. These optimized contours are obtained based on different contour forms for different load cases, objectives and optimization algorithms (q.v. [58]), while polynomial flexure hinges are not considered. In conclusion, these different recommendations show the difficulty of formulating simple and universal design guidelines. This is largely due to a strong dependence on the particular contour form, the investigated ratios of the basic hinge dimensions and the load case. Hence, general design graphs and design equations would be valuable in formulating a systematic and unified synthesis method for compliant mechanisms with concentrated compliance.

6.4 Angle-based synthesis method for individually shaped flexure hinges in a mechanism

As described in Section 6.1, the rigid-body replacement method is suitable for the synthesis of compliant mechanisms with optimal notch flexure hinges, especially for *path-generating mechanisms* providing high precision and large stroke. With a few exceptions (e.g. [3], [19], [21]), identical flexure hinges are generally used in a compliant mechanism. However, when compared with a separate hinge, the relative rotation angles φ^* of the desired motion in the rigid-body model are different for the incorporated hinges in most cases.

Due to the different rotation angles, different flexure hinge contours are also required. Because the deflection angle φ of each flexure hinge is approximately equal to the relative rotation angles φ^* in the rigid-body model (cf. Fig. 6.8), an *angle-based* and goal-oriented four-step *synthesis* method for using individually shaped notch flexure hinges in one compliant mechanism can be applied [58], see Fig. 6.4.

This synthesis method differs from existing approaches (e.g. [4], [48], [94]) especially with regards to the specific, angle-based and goal-oriented geometric design of each flexure hinge during synthesis. In previous literature, this important design step has been rarely investigated in detail and, where mentioned, has been limited to special contour forms, such as the corner-filleted flexure hinge [77], while the dimensional hinge ratios are varied. It is, however, also possible to vary or optimize the contour form for given hinge dimensions [62], for which design graphs are proposed in Section 6.4.1. Furthermore, both the contour form with its specific parameters and the basic hinge dimensions can be considered as free design parameters, for which design equations may be used as described in Section 6.4.2.

The basis of the geometric flexure hinge design is an investigation of a separate notch hinge, which is fixed at one end (cf. Fig. 6.3). A given moment **M** or direction-constant transverse force **F** leads to an ideal plane angular deflection of the free end with an angle of rotation φ. Two groups of geometric design parameters exert an influence on the properties of the notch flexure hinge:

- the basic hinge dimensions or link dimensions l, L, h, H, w, W, and
- the hinge contour height function $h_c(x)$.

The variable *hinge contour* height $h_c(x)$ is defined by the chosen notch geometry. Since not only the notch segment undergoes deformation ([108], [123]), the flexure hinge must be modeled with little segments of the both adjacent links always.

Due to an intuitive synthesis, the following assumptions were made for the notch hinge:

- the hinge contour is symmetric;
- the hinge contour is continuously differentiable and not undercut;
- the hinge has the minimum notch height h at its mid-point (at $x = 0$);
- the hinge cross-section is rectangular (so called prismatic notch flexure hinge).

Furthermore, the total height H, which also represents the link height in the compliant mechanism, as well as the total length L are chosen to be constant due to a unified synthesis process. Thus, the distance of the acting load to the middle of the hinge is always half the length of L, ensuring comparability.

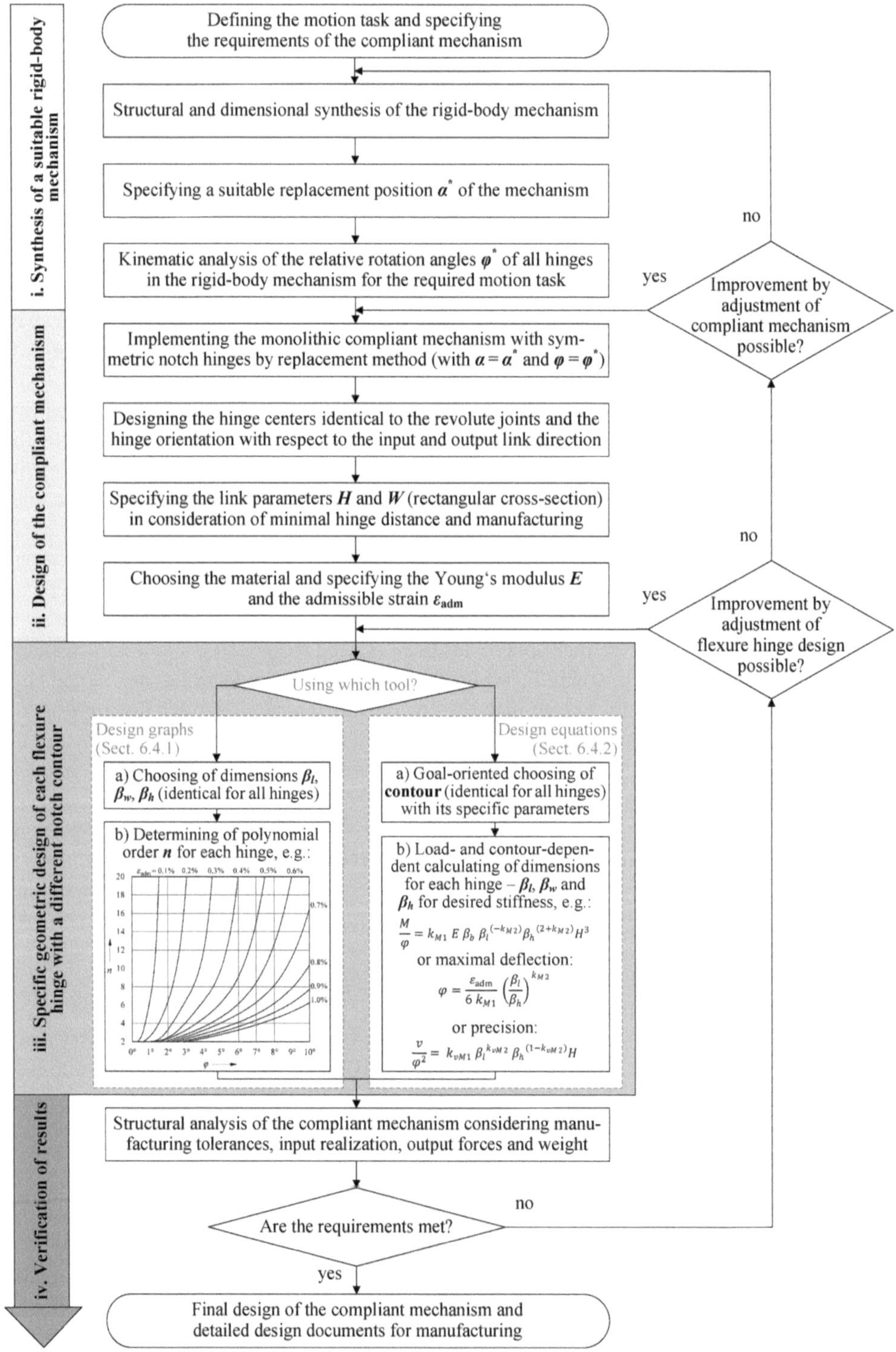

Fig. 6.4: Angle-based and goal-oriented synthesis of a compliant mechanism with individually shaped notch flexure hinges using the rigid-body replacement approach

The hinge length l, the minimum hinge height h, and the hinge width w can be varied within the design ranges according to the introduced dimensionless ratios β_l, β_h, and β_w with

$$\beta_l = \frac{l}{H}, \quad \beta_h = \frac{h}{H}, \quad \beta_w = \frac{w}{H}. \tag{6.1}$$

Applying the presented synthesis method, four notch flexure hinges can be used, three with usual contours and one with a novel variable polynomial contour (see Table 6.3):

- the hinge with a *semicircular notch contour*,
- the hinge with a *corner-filleted notch contour* with a general stress-optimal fillet radius,
- the hinge with a *semielliptical notch contour*,
- the hinge with a *polynomial notch contour* with an arbitrary adjustable order n (Section 6.4.1) or with six different typical orders for n (Section 6.4.2).

Table 6.3: Hinge contours and height functions for the four notch flexure hinges considered for synthesis (the hinges are shown for $\beta_l = 1$ and $\beta_h = 0.1$)

Hinge contour	Function $h_c(x)$ and restrictions
Semicircular 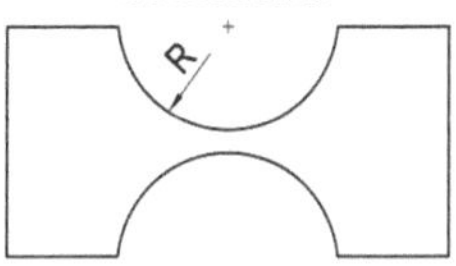	$h_c(x) = h + 2R - 2\sqrt{R^2 - x^2}$, with $R = 0.5l$
Corner-filleted 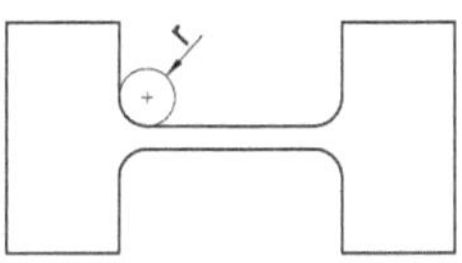	$h_c(x) = \begin{cases} h + 2r - 2\sqrt{r^2 - \left(x + \frac{l}{2} - r\right)^2}, & -\frac{l}{2} \le x < -\frac{l}{2} + r, \\ h, & -\frac{l}{2} + r \le x \le \frac{l}{2} - r, \\ h + 2r - 2\sqrt{r^2 - \left(x - \frac{l}{2} + r\right)^2}, & \frac{l}{2} - r < x \le \frac{l}{2}, \end{cases}$ with $r = 0.1l$
Semielliptical 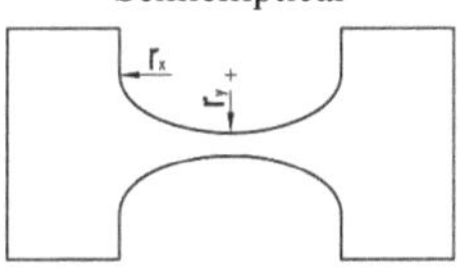	$h_c(x) = h + 2r_y\left(1 - \sqrt{1 - \frac{x^2}{r_x^2}}\right)$, with $r_x = 0.5l$ and $r_y = 0.25l$
Polynomial 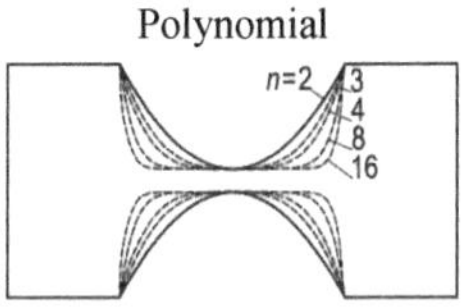	$h_c(x) = h + \frac{(H-h)}{\left(\frac{l}{2}\right)^n}\lvert x\rvert^n$, with $n \in \mathbb{N}$, $n \ge 2$

Typical or suitable parameter values, such as the stress-optimal fillet radius described in Section 6.3, have already been considered for contour modeling. The four contour height functions $h_c(x)$ are given in Table 6.3. Depending on the polynomial order n, the latter hinge contour represents a wide spectrum of different notch hinges, while five examples for a typical order are depicted. Based on FEM simulations, the influence of the order n on the hinge properties has been investigated. For a given deflection angle φ it has been shown that 16th-order polynomial contours lead to low stress values comparable to those of corner-filleted contours [60]. Furthermore, it was found that 4th-order polynomial contours allow for both a precise rotation and low stresses in general [58]. In addition, it is quite possible to directly optimize the order n in a compliant mechanism with different polynomial hinges [62].

6.4.1 Design of polynomial flexure hinges with variable order using design graphs

The detailed synthesis step of designing different polynomial flexure hinges with an angle-dependent order n using *design graphs* is shown in Fig. 6.5.

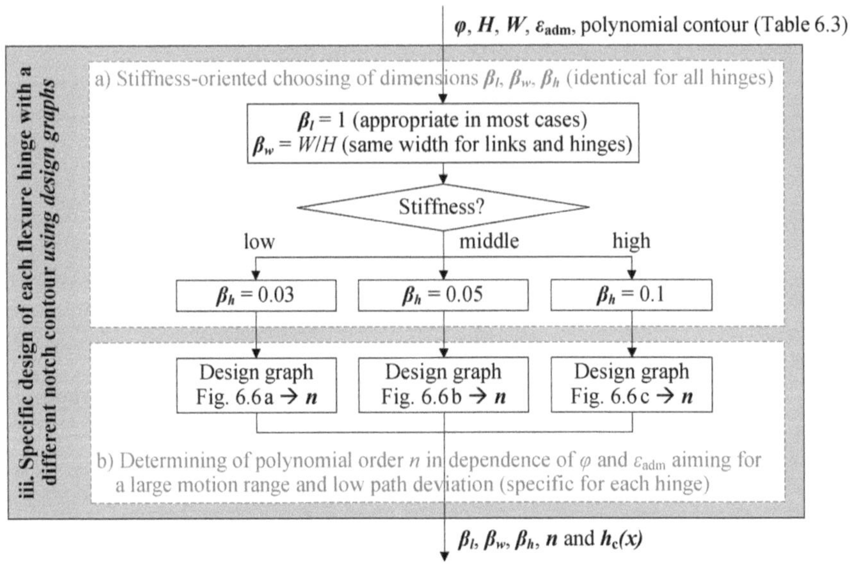

Fig. 6.5: Synthesis step of the angle-based geometric design of each notch flexure hinge with an individual polynomial contour using design graphs given for stiffness-specific dimensions

Based on the first two synthesis steps, the following mechanism parameters are specified and required (cf. Fig. 6.4):

- the link height H,
- the link width W,
- the maximum admissible elastic strain ε_{adm} of the selected material and
- in particular, the relative rotation angles φ of all incorporated hinges.

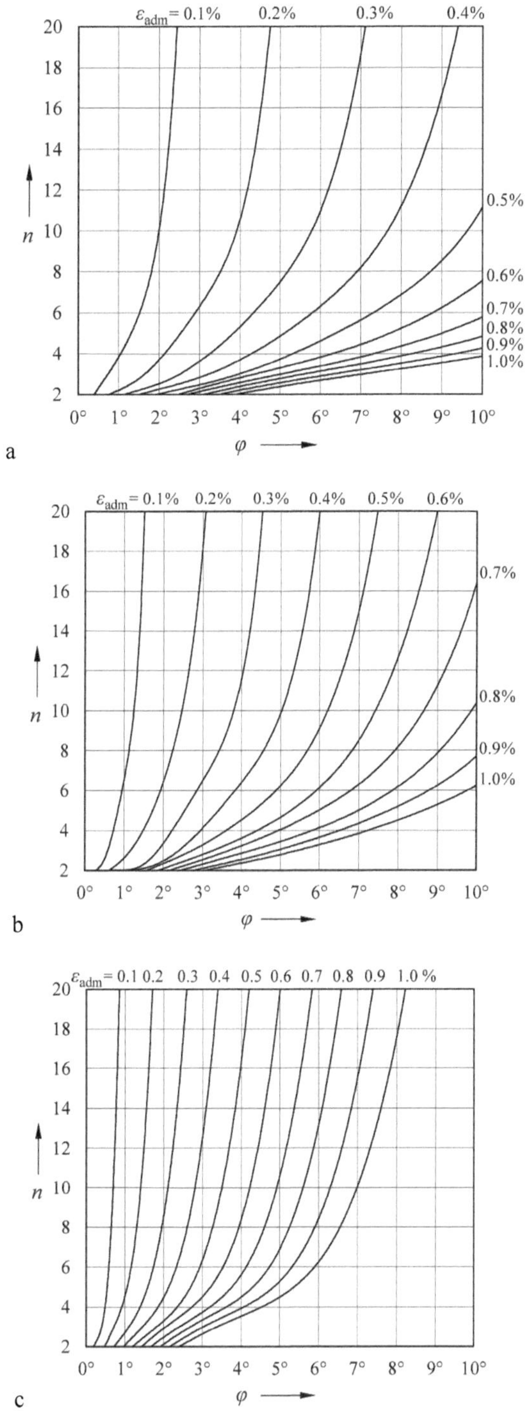

Fig. 6.6: FEM-based design graphs to determine the order n of each polynomial flexure hinge, depending on the rotation angle φ and the admissible material strain ε_{adm} for different hinge dimensions (odd numbers are also possible); a – for $\beta_l = 1$, $\beta_h = 0.03$; b – for $\beta_l = 1$, $\beta_h = 0.05$; c – for $\beta_l = 1$, $\beta_h = 0.1$

In the next step, first the hinge dimensions β_l and β_w must be specified as suggested, while the height dimension β_h must be selected according to the required hinge or mechanism stiffness, as a sum of the separate stiffnesses. Subsequently, all polynomial flexure hinges should initially have a low order (e.g. quadratic), in order to ensure high path accuracy of the compliant mechanism based on the rigid-body model. Accordingly, a higher polynomial order n must be determined for each hinge, using the related design graphs mentioned in Fig. 6.6. Thus, depending on φ, the maximum strain can be reduced to the value of ε_{adm}, with regard to the required large motion range of the mechanism. The design graphs presented here were developed based on FEM simulations in [58], and allow for the determination of n also with an odd number.

Hence, as result of the third synthesis step the following flexure hinge parameters are specified in the same way for each hinge:

- the hinge length ratio β_l,
- the link width ratio β_w and
- the link height ratio β_h.

Furthermore, the following parameters are specified individually for each flexure hinge:

- the polynomial order n and thus
- the final polynomial contour height $h_c(x)$, see Table 6.3.

6.4.2 Design of various flexure hinges with variable dimensions using design equations

The detailed synthesis design step that is used for usual or polynomial flexure hinges using contour-independent *design equations* is shown in Fig. 6.7.

As result of the first two synthesis steps, the following mechanism parameters are specified and required in this case (cf. Fig. 6.4):

- the link height H,
- the link width W,
- the YOUNG's modulus E and admissible elastic strain ε_{adm} of the selected material and
- in particular again, the relative rotation angles φ of all incorporated flexure hinges.

In the next step, first the hinge contour with its specific parameters must be selected from nine geometries according to the mechanism design goals (cf. Table 6.3). Subsequently, the hinge dimensions β_l and β_w can be specified as suggested, while β_h must be calculated with the design equations according to the most important performance criteria and depending on the estimated load case. The other hinge properties can, of course, also be verified to determine if any geometric parameters must be adjusted. If necessary, β_l and β_w can be varied, too.

Hence, as result of the third synthesis step, the following flexure hinge parameters are specified identically or differently for each hinge: the hinge ratios β_l, β_w and β_h, and the final contour height $h_c(x)$ with its specific parameter values (see Table 6.3).

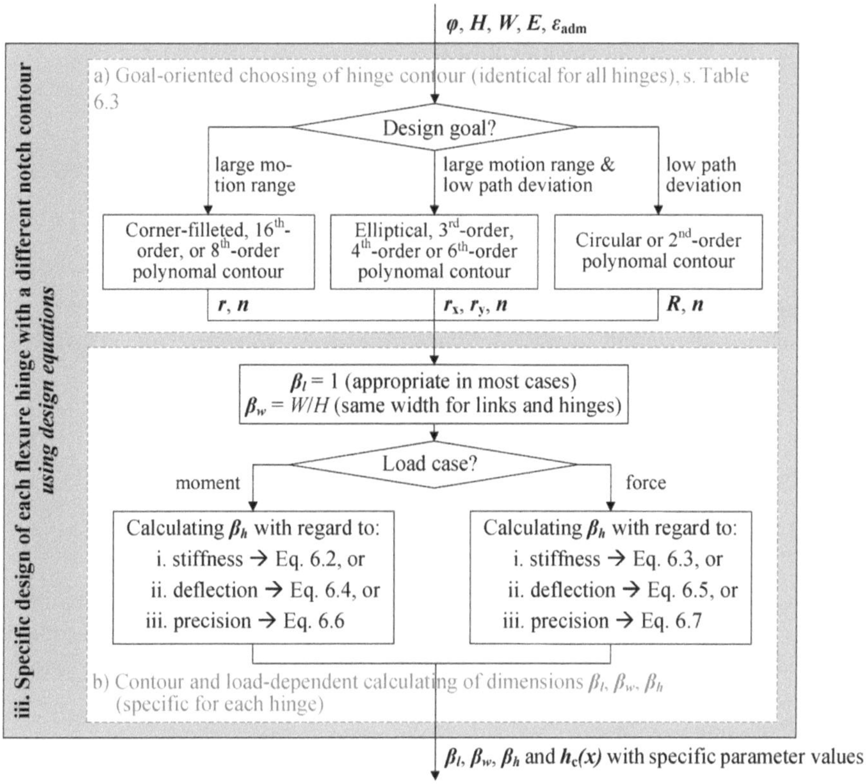

Fig. 6.7: Synthesis step of the angle-based geometric design of each notch flexure hinge with an individual commonly-used or polynomial contour using contour-independent design equations

The six contour-independent, closed-form design equations (6.2)-(6.7) for a flexure hinge with various notch contours are developed based on the analytical characterization due to the non-linear theory for modeling large deflections of the curved rod-like structures presented in Section 4.7.6. Three hinge performance criteria are considered: the bending stiffness, the maximum angular deflection and the rotational precision. These equations are developed for a moment or a transverse force load and are accurately valid for a rotation angle up to $\varphi = 5°$. The calculation of the results for larger angles is nevertheless possible. The load acts close to the hinge center at $L = 2H$ (free end), while only the elastic properties are almost independent from the value of this distance for the moment load.

In this case, the elasto-kinematic hinge properties are expressed as a power function with two coefficients in dimensionless form, on the basis of various analytical results derived numerically with MATLAB (q.v. [66]). The power function was chosen because its form is suitable for considering the principle dependence on the basic geometric parameters, which increases strictly monotonically for different contours. Furthermore, the power function offers great shape variability with a low number of coefficients that can be used to consider the contour form. SI units must be used for all parameters.

Bending Stiffness
For a moment load, the design equation for the *bending stiffness* is given as:

$$\frac{M}{\varphi} = k_{M1}\, E\, \beta_w\, \beta_l^{-k_{M2}}\, \beta_h^{2+k_{M2}}\, H^3 \tag{6.2}$$

and, taking into account a transverse force load, the force-specific bending stiffness is:

$$\frac{F}{\varphi} = k_{F1}\, E\, \beta_w\, \beta_l^{-k_{F2}}\, \beta_h^{2+k_{F2}}\, H^2. \tag{6.3}$$

Maximum angular deflection
For a moment load, the design equation for the *maximum angular deflection* is given as:

$$\left|\varphi_{\max}\right| = \frac{\varepsilon_{\text{adm}}}{6\,k_{M1}} \left(\frac{\beta_l}{\beta_h}\right)^{k_{M2}} \tag{6.4}$$

and, taking into account a transverse force load, the maximum angular deflection is:

$$\left|\varphi_{\max}\right| = \frac{\varepsilon_{\text{adm}}}{12\left(1-k_{\text{crit}}\right)k_{F1}} \left(\frac{\beta_l}{\beta_h}\right)^{k_{F2}}. \tag{6.5}$$

According to the theory, the maximum strain occurs independently of the contour in the hinge center for a moment, while an additional dimensionless factor k_{crit} has become necessary to consider the deviation of the critical strain location from the hinge center for a force.

Rotational Precision
For a moment load, the design equation for the *rotational axis shift* is given as:

$$\frac{v}{\varphi^2} = k_{vM1}\, \beta_l^{k_{vM2}}\, \beta_h^{1-k_{vM2}}\, H \tag{6.6}$$

and, taking into account a transverse force load, the rotational axis shift is:

$$\frac{v}{\varphi} = k_{vF1}\, \beta_l^{k_{vF2}}\, \beta_h^{2-k_{vF2}}\, H. \tag{6.7}$$

The rotational axis shift v was determined based on a model and using the fixed center approach (q.v. [66]). Therefore, only the deformation of one arbitrary point at the free end of the flexure hinge and the bending angle must be known. According to the analytical results for $\varphi \leq 5°$, the axis shift-angle-behavior is considered non-linear (quadratic) for the moment load, and nearly linear for the force load.

Contour-specific coefficients
The contour-specific coefficients of all six design equations are given in Table 6.4 for nine notch flexure hinge contours and an appropriate parameter range of the hinge dimensions, the hinge length ratio β_l ($0.5 \leq \beta_l \leq 1.5$) and the hinge height ratio β_h ($0.03 \leq \beta_h \leq 0.1$).

The coefficients of the power functions are determined with MATLAB, based on a fitting procedure in order to attain the smallest maximum error over all calculated result points. This approach is explained in [66] in detail. The relative discrepancy errors between the design equations results and the analytical results, as well as a comparison with FEM results, are mentioned in [65]. According to the theory, the accuracy of the results is nearly independent of the parameter range for the hinge width ratio β_w.

Table 6.4: Coefficients for design eq. (6.2)-(6.7) depending on the hinge contour, valid for appropriate dimensions ($0.5 \leq \beta_l \leq 1.5$, $0.03 \leq \beta_h \leq 0.1$, β_w arbitrary) and $\varphi \leq 5°$ (the hinges are shown for $\beta_l = 1$ and $\beta_h = 0.03$)

Hinge contour	k_{M1} [10^{-3}]	k_{M2}	k_{F1} [10^{-2}]	k_{F2}	k_{crit}	$k_{\nu M1}$ [10^{-3}]	$k_{\nu M2}$	$k_{\nu F1}$ [10^{-2}]	$k_{\nu F2}$
Semicircular 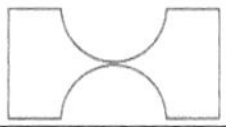	107.90	0.52	10.55	0.51	0.5	99.85	0.52	19.12	0.94
Corner-filleted 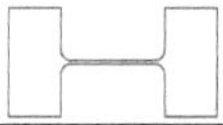	83.95	0.96	8.41	0.96	$0.5 - 0.2\beta_l$	85.76	0.95	9.20	1.89
Semielliptical 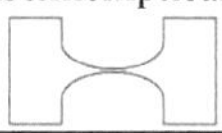	82.50	0.54	8.27	0.54	0.5	114.35	0.57	18.21	1.14
2nd-ord. polynomial 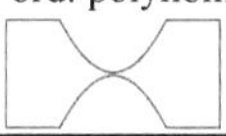	133.00	0.48	13.32	0.48	$0.4\beta_l^{-0.005}\beta_h^{-0.075}$	80.27	0.47	12.88	0.88
3rd-ord. polynomial 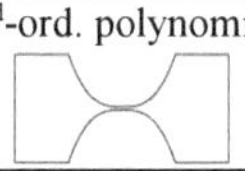	120.23	0.65	12.04	0.65	$0.4\beta_l^{-0.035}\beta_h^{-0.064}$	72.25	0.64	6.67	1.29
4th-ord. polynomial 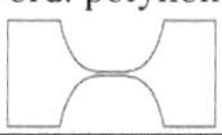	112.07	0.74	11.22	0.74	$0.4\beta_l^{-0.081}\beta_h^{-0.048}$	71.32	0.73	6.24	1.47
6th-ord. polynomial 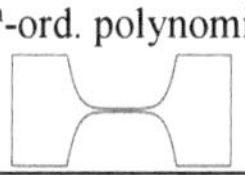	103.02	0.83	10.32	0.83	$0.4\beta_l^{-0.212}\beta_h^{-0.015}$	72.53	0.82	6.35	1.64
8th-ord. polynomial 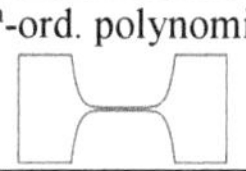	98.24	0.87	9.84	0.87	$0.4\beta_l^{-0.283}\beta_h^{-0.01}$	74.00	0.87	6.59	1.73
16th-ord. polynomial 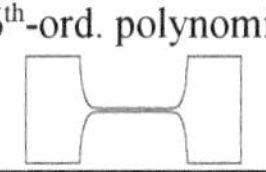	90.82	0.93	9.10	0.93	$0.4\beta_l^{-0.463}\beta_h^{-0.07}$	77.63	0.93	7.25	1.87

Existing design equations from literature (e.g. [16], [68], [78], [83], [98], [108]) are mostly characterized by a long expression and complex structural form. Furthermore, they are often only valid for a specific group of flexure hinge contours, and have not been developed for all three rotational properties. Hence, with regard to an accelerated and unified synthesis of compliant mechanisms, the general design equations presented here are concise and thus advantageous. With only two coefficients, their structural form is simple, contour-independent and, with respect to the link height H as scaling factor, dimensionless. *Geometric scaling* is a further suitable design approach to adjust an initial compliant mechanism [61].

6.5 Examples of compliant mechanisms with polynomial flexure hinge design

In this section, the synthesis method suggested in Section 6.4 is applied to a high-precise and large-stroke compliant four-bar *guidance mechanism* to explore the angle-based approach of individual and optimal polynomial flexure hinge design. In addition, twelve further linear path-generating mechanisms, developed using this synthesis method, are presented.

6.5.1 Angle-based synthesis of a high-precision and large-stroke straight-line mechanism with different polynomial hinges

A four-bar mechanism after HOECKEN used to guide the coupler point P on an approximated rectilinear path [47] is chosen as an example. The rigid-body model and its compliant counterpart are shown in Fig. 6.8 in both initial and deflected positions (cf. Fig. 6.1). The resulting compliant mechanism with different polynomial hinges is exemplarily synthesized based on the design graph approach presented in Section 6.4.1 (cf. Fig. 6.4 and Fig. 6.5).

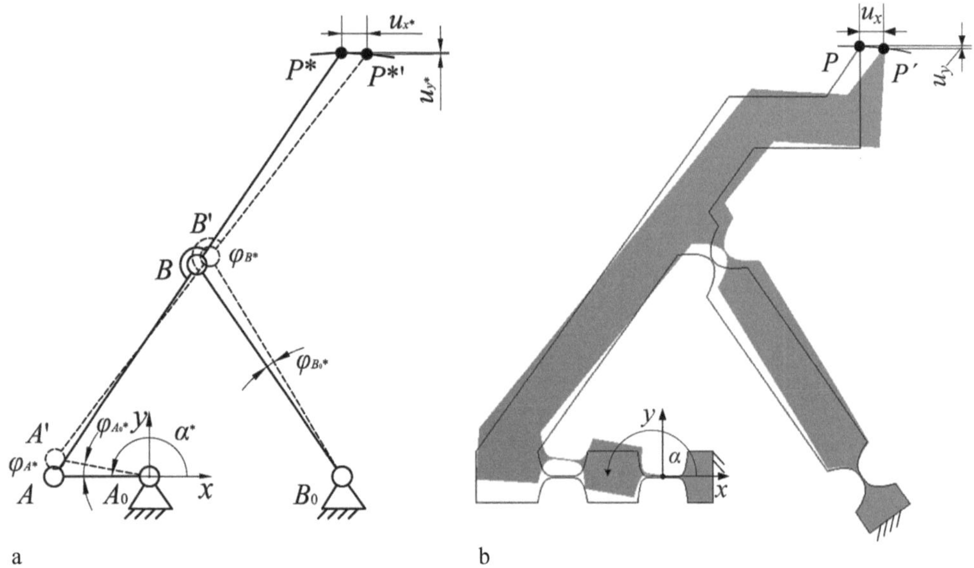

Fig. 6.8: Four-bar straight-line mechanism for an approximated rectilinear guidance of the coupler point in initial (P) and deflected (P') positions due to an input displacement $u_{x*} = u_x = 5$ mm; a – rigid-body model; b – compliant mechanism with different polynomial flexure hinges ($\beta_l = 1$, $\beta_h = 0.03$)

i. Synthesis of a suitable rigid-body mechanism

- The required motion task is a straight-line guidance of the coupler point P of ±5 mm.
- An asymmetric four-bar mechanism with a closed symmetric four-point contact tangent path for one complete crank rotation is selected. The mechanism is based on an isosceles crank-and-rocker linkage. According to [101], p. 38, the following link length relations are considered with respect to a small straight-line deviation:

$$\overline{A_0B_0} = 2\,\overline{A_0A},\ \ \overline{AB} = \overline{B_0B} = \overline{BP} = \frac{3}{2}\sqrt{3}\,\overline{A_0A} \ \text{and}\ \overline{AP} = 3\sqrt{3}\,\overline{A_0A}. \tag{6.8}$$

- The replacement mechanism position $\alpha^* = \alpha = 180°$ is chosen.
- Based on a kinematic analysis, the four relative rotation angles of each hinge (6.9) and the relevant straight-line deviation of the rigid-body model (6.10) are determined for the input motion $u_{x*} = 5$ mm and the crank length $\overline{A_0A} = 20\,\text{mm}$.

$$\varphi_{A_0*} = 10.13°,\ \varphi_{A*} = 6.9°,\ \varphi_{B*} = 0.28° \ \text{and}\ \varphi_{B_0*} = 3.51° \tag{6.9}$$

$$u_{y*} = -0.38\,\mu\text{m} \tag{6.10}$$

ii. Design of the compliant mechanism

- The compliant mechanism is designed as described in Fig. 6.4.
- The link parameters are specified with $H = 10$ mm and $W = H$.
- The material is specified as aluminum AW 7075 with $E = 72$ GPa and $\varepsilon_{adm} = 0.5$ %.

iii. Design of each hinge with an individual polynomial contour using design graphs

- The hinge length and the hinge width ratio are specified as $\beta_l = \beta_w = 1$.
- The hinge height ratio is chosen as $\beta_h = 0.03$ with regard to a low bending stiffness.
- According to the design graph in Fig. 6.9 (cf. Fig. 6.6 a), the polynomial orders of the four hinges are determined with the hinge angles φ that are approximately equal to the angles φ^*, see equation (6.11). An odd order of n can also be achieved, if a quarter hinge is modeled first, which must be mirrored twice in the CAD model afterwards. Furthermore, if the polynomial contour is extended to a power function contour [44], rational numbers are possible with regard to a more accurate determination of each order or exponent n.

$$n_{A_0} = 11, \; n_A = 6, \; n_B = 2 \text{ and } n_{B_0} = 3 \tag{6.11}$$

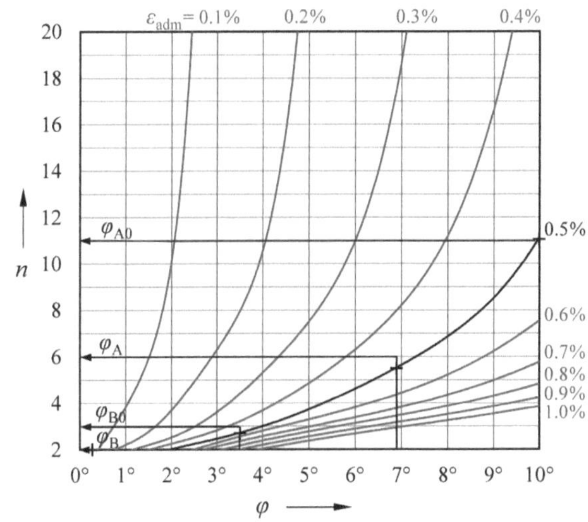

Fig. 6.9: Illustration of using the design graph of Fig. 6.6 a to determine the polynomial hinge orders

- The resulting compliant mechanism with optimal and different polynomial contours is shown in Fig. 6.10 c, with a comparison of mechanisms using common contours.

iv. FEM-based verification of results

- The mechanism properties are calculated using a quasi-static geometrically non-linear FEM simulation. The straight-line deviation which corresponds with the path of motion is shown in Fig. 6.11, and further results are mentioned in Table 6.5.

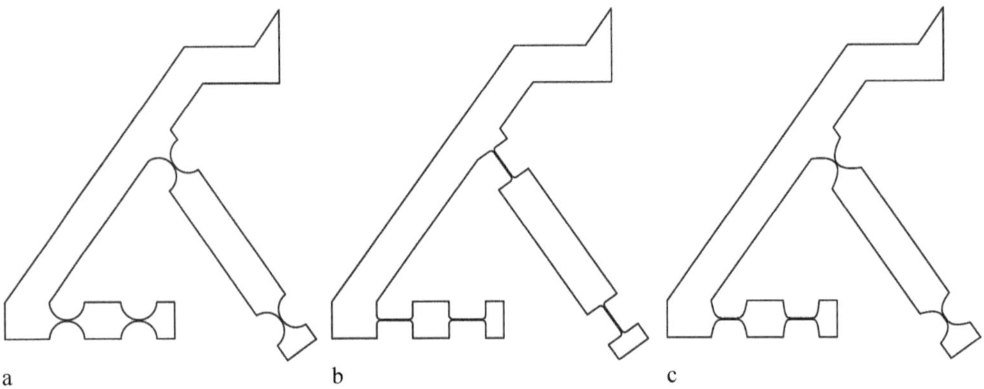

Fig. 6.10: Geometric designs of the compliant four-bar mechanism ($\beta_l = 1$, $\beta_h = 0.03$); a – with identical semicircular hinges ($R = 5$ mm); b – with identical corner-filleted hinges ($r = 1$ mm); c – with different polynomial hinges using the graph-based synthesis method presented in Sect. 6.4.1 ($n_{A0} = 11$, $n_A = 6$, $n_B = 2$, $n_{B0} = 3$)

From analyzing the path of motion, it can be seen that all three compliant four-bar mechanisms exhibit the desired small straight-line deviation in the micrometer range for the input displacement $u_x = \pm 5$ mm. As in the single hinge results, the mechanism with semicircular hinges provides the smallest straight-line deviation. Due to the very small deviation of the rigid-body model, less than one micrometer, the path deviations of all compliant mechanisms qualitatively and quantitatively show the same behavior as the straight-line deviations.

With regard to the maximum *admissible strain*, the required input or output stroke cannot be realized when using the mechanism with semicircular hinges. In contrast, the required stroke is possible when using the mechanism with corner-filleted hinges, and as expected, also with the synthesized mechanism with individually shaped polynomial hinges in approximation. Furthermore, the sensitivity to vibrations is reduced due to a higher stiffness.

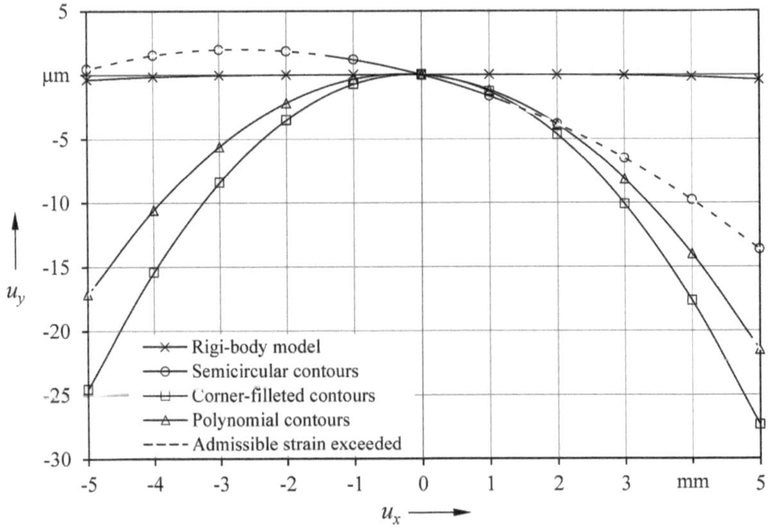

Fig. 6.11: FEM simulation results for the motion path of the coupler point P of the four-bar straight-line mechanisms considered here (the curves are drawn in dashed lines from the input displacement value at which the admissible strain $\varepsilon_{adm} = 0.5$ % is exceeded)

Table 6.5: FEM results for the compliant four-bar straight-line mechanisms with commonly-used semicircular and corner-filleted flexure hinges, and with different polynomial flexure hinges (input $u_x = 5$ mm)

Hinge contours	Straight-line deviation u_y [µm]	Path deviation $\lvert u_y - u_{y*} \rvert$ [µm]	Deflection force F_x [N]	Max. von-Mises strain ε_{max} [%]	Admissible input stroke u_{xadm} [mm]
Identical semicircular 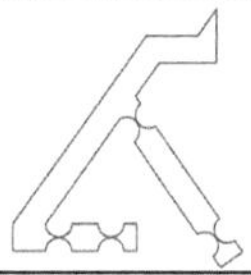	−13.63	13.25	11.90	1.89	1.32
Identical corner-filleted 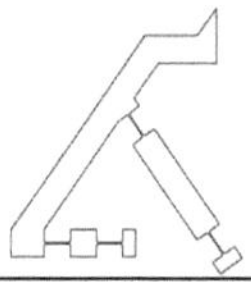	−27.36	26.98	1.88	0.41	6.10
Different polynomial 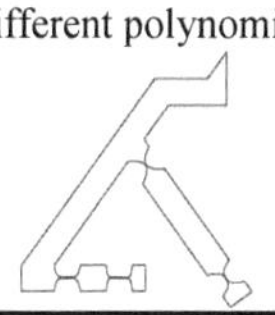	−21.50	21.12	3.38	0.51	4.90

Thus, considering the two design goals of high precision and large stroke, the applied graph-based synthesis method (cf. Section 6.4.1) allows for the accelerated development of a compliant mechanism with optimal polynomial flexure hinge design. The dimensional restrictions can be avoided by using the presented design equations (cf. Section 6.4.2). Taking into account a specific required stiffness or deflection force of the compliant mechanism, a more accurate design with additional weighting of the design goals is also possible with the equation-based synthesis method for differently and individually designed hinges.

Furthermore, computational *design tools* may be used for the comprehensive analysis and synthesis of various notch flexure hinges, such as the indigenously developed software *detasFLEX* [44], which is based on the non-linear modeling approach presented in this book.

6.5.2 Further examples of synthesized mechanisms

Additional compliant planar *path-generating mechanisms* with individually shaped polynomial hinges are shown in Table 6.6 and Table 6.7, which have been developed to investigate the applicability of the angle-based synthesis method, especially for mechanisms with a higher link number.

Six rectilinear point and six rectilinear plane-*guidance mechanisms* are shown, for which the straight-line deviation is in the low micrometer range given a large stroke of several millimeters or even centimeters. The output motion is symbolized by an arrow. For the plane-guidance mechanisms the orientation of the whole guided link remains constant. Due to double hinges with same axes positions, some mechanisms are realized in a non-planar way.

In conclusion, the suggested angle-based synthesis method can be applied to fully and partially compliant mechanisms, and offers new possibilities in the field of optimal compliant mechanism design due to the use of different notch flexure hinges in one mechanism.

Table 6.6: Compliant mechanisms with different polynomial hinges for linear guidance of a point ($\beta_l = 1, \beta_h = 0.03$)

Mechanism, reference	Mechanism structure	Hinges	Polynomial orders n	Compliant mechanism example
Isosceles slider crank mechanism, [62]	Partially compliant	3	4, 8, 4	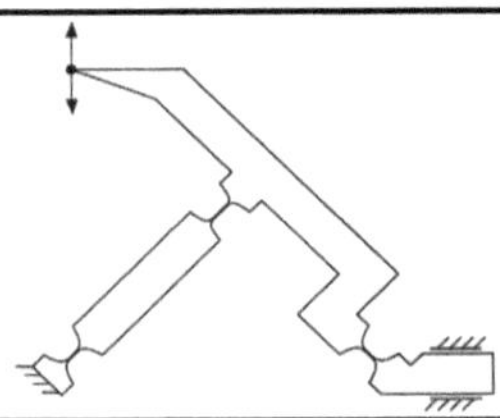
EVANS mechanism, [35]	Fully compliant	4	3, 9, 16, 5	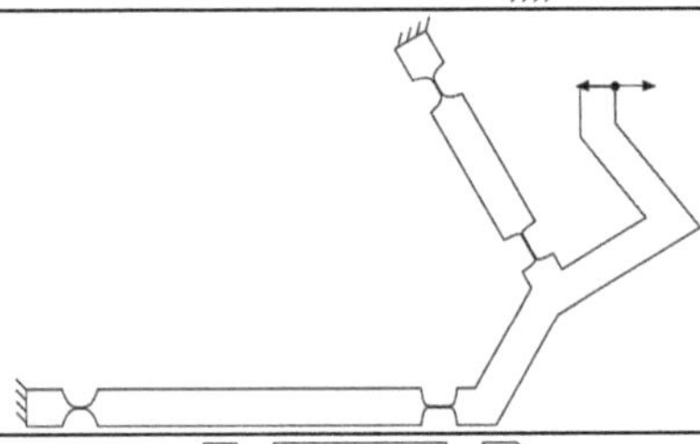
ROBERTS mechanism, [35]	Fully compliant	4	3, 11, 11, 3	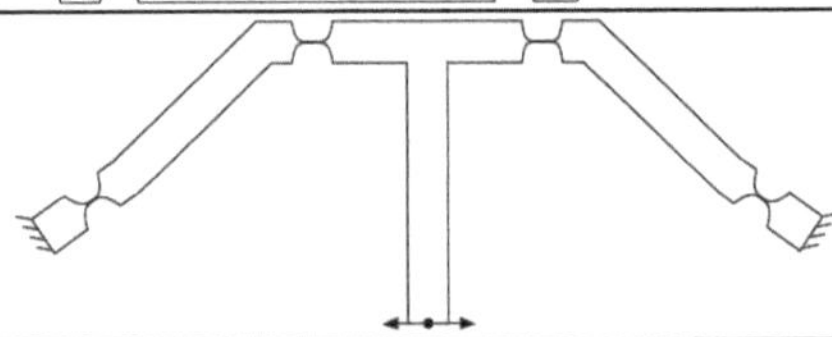
WATT mechanism, [58]	Fully compliant	4	5, 14, 14, 5	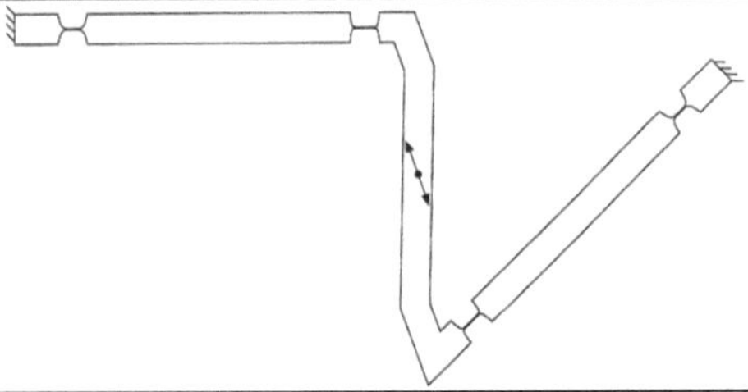
Pantograph-based mechanism, [58]	Partially compliant	6	4 (2 hinges), 8 (4 hinges)	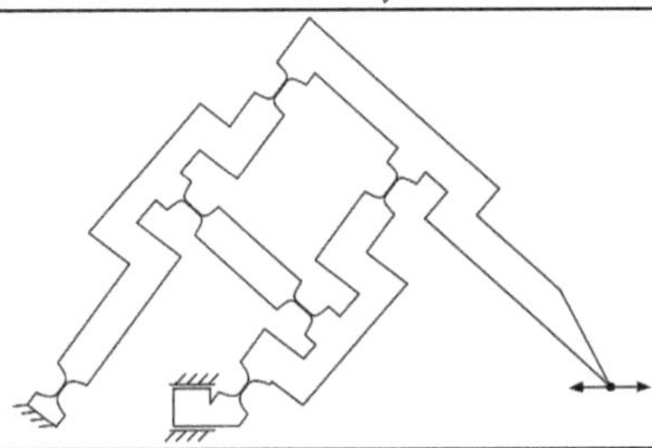
PEAUCELLIER mechanism, [53]	Fully compliant	10	2 (6 hinges), 3 (3 hinges), 6 (1 hinge)	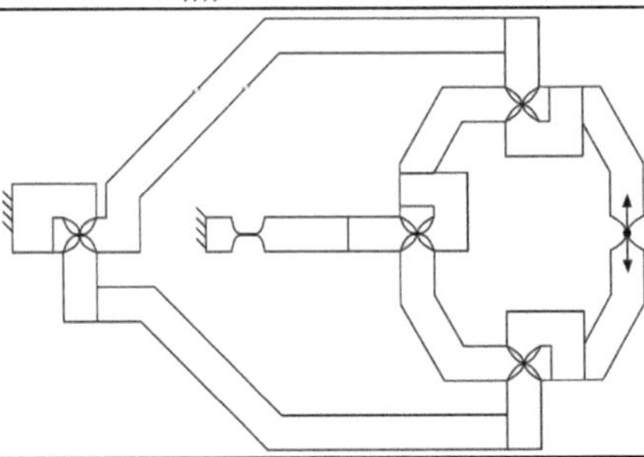

Table 6.7: Compliant mechanisms with different polynomial hinges for linear guidance of a plane (β_l=1, β_h=0.03)

Mechanism, reference	Mechanism structure	Hinges	Polynomial orders n	Compliant mechanism example
Parallel crank mechanism, [34]	Fully compliant	4	6 (all hinges)	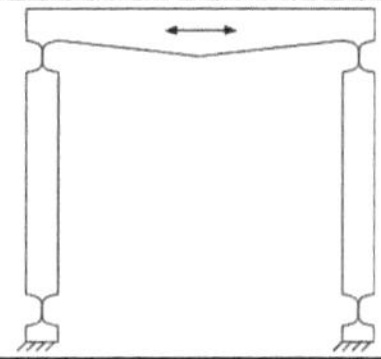
Gripper based on a four-bar mechanism (half gripper), [63]	Fully compliant	4	2, 2, 5, 5	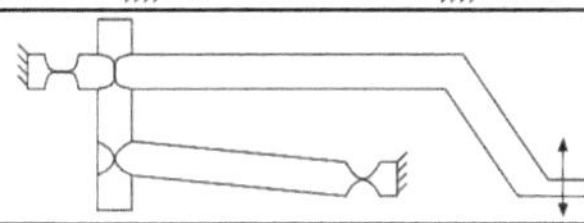
Gripper based on a parallel crank mechan. (half gripper), [63]	Fully compliant	7	2 (2 hinges), 4 (5 hinges)	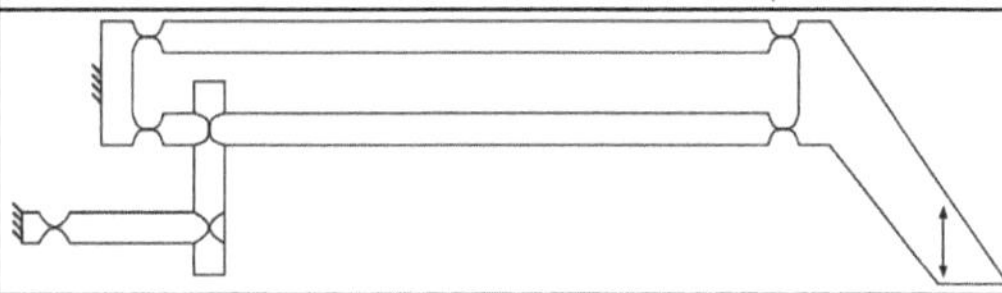
10-Hinge mechanism based on two serially combined parallel crank mechanisms, [33]	Fully compliant	10	4 (all hinges)	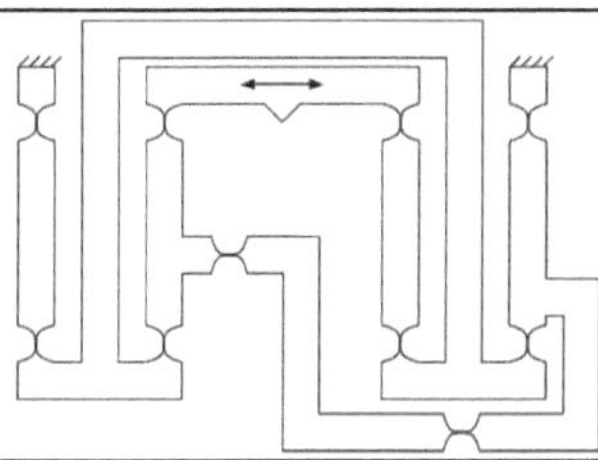
10-Hinge mechanism based on ROBERTS mechanism, [53]	Fully compliant	10	2 (4 hinges), 3 (3 hinges), 4 (3 hinges)	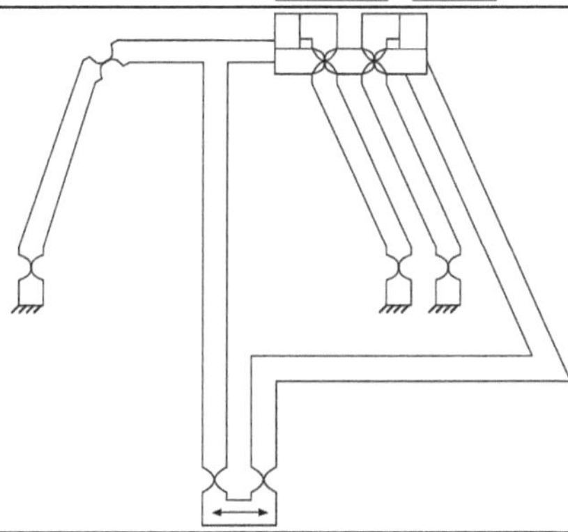
12-Hinge mechanism based on enhanced pantograph, rigid-body model after [46]	Partially compliant	12	2 (3 hinges), 5 (5 hinges), 7 (4 hinges)	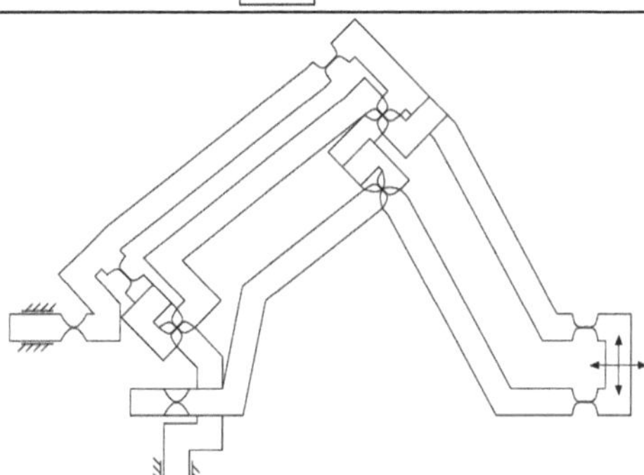

References

[1] Baule, B.: Die Mathematik des Naturforschers und Ingenieurs, Band VII: Differentialgeometrie, Leipzig, Hirzel Verlag, Lizenz-Nr. 267-245/48/65 ES19 B3, 12–14, 1965

[2] Bernardoni, P., Bidaud, P., Bidard, C., Gosselin, F.: A new compliant mechanism design methodology based on flexible building blocks, In: Proceedings of the SPIE 2004 Conference, Smart Structures and Materials 2004, San Diego, USA, 244–254, doi:10.1117/12.539498, 2004

[3] Beroz, J., Awtar, S., John Hart, A.: Extensible-Link Kinematic Model for Characterizing and Optimizing Compliant Mechanism Motion, Journal of Mechanical Design, 136, 031008, doi:10.1115/1.4026269, 2014

[4] Bharanidaran, R., Ramesh, T.: A numerical approach to design a compliant based Microgripper with integrated force seninsing jaw, International Journal of Mechanics, 7, 128–135, 2013

[5] Bi, S., Zhao, H., Yu, J.: Modeling of a Cartwheel Flexural Pivot, Journal of Mechanical Design, 131, 61010, doi:10.1115/1.3125204, 2009

[6] Bi, S., Zhao, S., Zhu, X.: Dimensionless design graphs for three types of annulus-shaped flexure hinges, Precision Engineering, 34, 659–667, doi:10.1016/j.precisioneng.2010.01.002, 2010

[7] Bicchi, A., Tonietti A.: Fast and “soft-arm” tactics, IEEE Robotics and Automation Magazine, 11, No. 2, 22–33, doi:10.1109/MRA.2004.1310939, 2004

[8] Bögelsack, G.: Nachgiebige Mehanismen in miniaturisierten Bewegungssystemen, In: 9th World Congress on Theory of Machines and Mechanisms, Milano, Italy, 3101–3104, 1995

[9] Bögelsack, G.: On Fluidmechanical Compliant Actuators, 19th Working Meeting of IFToMM, Kaunas-Technologija, Lithuania, 51–56, 2000

[10] Bona, F. De, Munteanu, M. G.: Optimized Flexural Hinges for Compliant Micromechanisms, Analog Integrated Circuits and Signal Processing, 44, 163–174, doi:10.1007/s10470-005-2597-7, 2005

[11] Böttcher, F., Pfefferkorn, H.: Gelenkig bewegliche Verbindungen mit besonderen Eigenschaften, In: Proceedings of the 44th International Scientific Colloquium, Ilmenau, Germany, 240–245, 1999

[12] Brouwer, D. M., Wiersma, H., Boer, S., Aarts, R.: Large-stroke spatial hinge flexures: A method for optimising the geometry of flexure hinges, aimed at maximising supporting stiffnesses, Mikroniek, 53, 24–30, 2013

[13] Cannon, J. R., Howell, L. L.: A compliant contact-aided revolute joint, Mechanism and Machine Theory, 40, 1273–1293, doi:10.1016/j.mechmachtheory.2005.01.011, 2005

[14] Chaykina, A., Griebel, S., Zentner, L.: Nachgiebiges Sensorsystem zur Ermittlung von Scherkräften, In: 10. Kolloquium Getriebetechnik, 2013.09.11–13, Ilmenau, Univ.-Verl., 419–436, ISSN 2194-9476, 2013

[15] Chaykina, A., Griebel, S., Zentner, L.: Richtungsabhängiger Berührungssensor zur Sensorisierung von nachgiebigen Mechanismen, Ilmenau, Univ.-Verl., Berichte der Ilmenauer Mechanismentechnik 1, 79–90, ISSN 2194-9476, 2012

https://doi.org/10.1515/9783110479744-167

[16] Chen, G., Wang, J., Liu, X.: Generalized Equations for Estimating Stress Concentration Factors of Various Notch Flexure Hinges, Journal of Mechanical Design, 136, 031009, doi:10.1115/1.4026265, 2014

[17] Chen, G.-M., Jia, J.-Y., Li, Z.-W.: Right-circular Corner-filleted Flexure Hinges, In: Proceedings of the 2005 IEEE International Conference on Automation Science and Engineering, Edmonton, Canada, 249–253, doi:10.1109/COASE.2005.1506777, 2005

[18] Christen, G., Pfefferkorn, H.: Mehr Beweglichkeit: Nachgiebige Mechanismen eignen sich als elastische Getriebe, MM – Das Industrie-Magazin, 37, 2002

[19] Christen, G., Pfefferkorn, H.: Nachgiebige Mechanismen: Aufbau, Gestaltung, Dimensionierung und experimentelle Untersuchung, In: VDI-Berichte Nr. 1423 1998, VDI-Getriebetagung, Kassel, Germany, 309–329, 1998

[20] Christen, G., Pfefferkorn, H.: Zum Bewegungsverhalten nachgiebiger Mechanismen, Wissenschaftliche Zeitschrift der Universität Dresden, 50, 53–58, 2001

[21] Clark, L., Shirinzadeh, B., Zhong, Y., Tian, Y., Zhang, D.: Design and analysis of a compact flexure-based precision pure rotation stage without actuator redundancy, Mechanism and Machine Theory, 105, 129–144, doi:10.1016/j.mechmachtheory.2016.06.017, 2016

[22] Cosandier, F., Eichenberger, A., Baumann, H., Jeckelmann, B., Bonny, M., Chatagny, V., Clavel, R.: Development and integration of high straightness flexure guiding mechanisms dedicated to the METAS watt balance Mark II, Metrologia, 51, 88–95, doi:10.1088/0026-1394/51/2/S88, 2014

[23] Desrochers, S.: Optimum design of simplical uniaxial accelerometers, Master‘s thesis, Montréal, McGill University, 2008

[24] Diener, L. E.: FEM-basierte Untersuchung verschiedener Festkörpergelenktypen und Vergleich ihrer Anwendung in hochpräzisen nachgiebigen Mechanismen, Master‘s thesis, Ilmenau, Technische Universität Ilmenau, 2017

[25] Dirksen, F., Lammering, R.: On mechanical properties of planar flexure hinges of compliant mechanisms, Mechanical Sciences, 2, 109–117, doi:10.5194/ms-2-109-2011, 2011

[26] Du, Z., Yang, M., Dong, W.: Multi-objective optimization of a type of ellipse-parabola shaped superelastic flecure hinge, Mechanical Sciences, 7, 127–134, doi:10.5194/ms-7-127-2016, 2016

[27] Duarte, R., Howells, M. R., Hussain, Z., Lauritzen, T.: A Linear Motion Machine for Soft X-ray Interferometry, In: Proceedings of the SPIE 1997 Conference, San Diego, USA, doi:10.1117/12.279100, 1997

[28] Feierabend, M., Zentner, L.: Konzeption eines Mechanotherapie-Systems zur Rehabilitation der Handfunktionalität für den Einsatz in der medizinischen Trainingstherapie, In: 10. Kolloquium Getriebetechnik, 2013.09.11–13, Ilmenau, Univ.-Verl., 437–453, ISSN 2194-9476, 2013

[29] Fettig, H., Wylde, J., Hubbard, T., Kujath, M.: Simulation, dynamic testing and design of micromachined flexible joints, Journal of micromechanics and microengineering structures, devices and systems, 11, 209–216, doi:10.1088/0960-1317/11/3/308, 2001

[30] Filonenko-Boroditsch, M. M.: Festigkeitslehre, II, Berlin, Verlag Technik, 1952

[31] Frecker, M. I., Ananthasuresh, G. K., Nishiwaki, S., Kota, S.: Topological Synthesis of Compliant Mechanisms Using Multi-Criteria Optimization, Journal of Mechanical Design, 119, 238–245, doi:10.1115/1.2826242, 1997

[32] Goldfarb, M., Speich, J. E.: A Well-Behaved Revolute Flexure Joint for Compliant Mechanism Design, Journal of Mechanical Design, 121, 424–429, doi:10.1115/1.2829478, 1999

[33] Gräser, P., Linß, S., Harfensteller, F., Zentner, L., Theska, R.: Large stroke ultra-precision planar stage based on compliant mechanisms with polynomial flexure hinge design, In: Proceedings of the 17th International conference of the European Society for Precision Engineering and Nanotechnology (euspen), Hannover, Germany, 207–208, ISBN 978-0-9957751-0-7, 2017

[34] Gräser, P., Linß, S., Zentner, L., Theska, R.: Design and Experimental Characterization of a Flexure Hinge-Based Parallel Four-Bar Mechanism for Precision Guides, In: Microactuators and Micromechanisms, Zentner, L. et al. (Eds.), Mechanisms and Machine Science, 54, Cham, Springer International Publishing, 139–152, doi:10.1007/978-3-319-45387-3_13, 2017

[35] Gräser, P., Linß, S., Zentner, L., Theska, R.: Optimization of Compliant Mechanisms by Use of different Polynomial Flexure Hinge Contours, In: 3rd IAK, Interdisciplinary Applications of Kinematics, Lima, Peru, 2018

[36] Griebel, S., Feierabend, M., Bojtos, A., Zentner, L.: Kennlinien eines nachgiebigen fluidmechanischen Antriebes zur Erzeugung einer schraubenförmigen Bewegung – Vergleich Simulation und Messaufbau, In: 10. Kolloquium Getriebetechnik, 2013.09.11–13, Ilmenau, Univ.-Verl., 391–498, ISSN 2194-9476, 2013

[37] Griebel, S., Streng, A., Zentner, L.: Nachgiebiger Fluidantrieb zur Erzeugung einer nahezu exakten bidirektionalen Schraubenbewegung und dazugehöriges Verfahren, German Patent DE102011104026B4, filed 8. Jun. 2011, issued 11. Apr. 2013

[38] Griebel, S., Zentner, L., Böhm, V., Haueisen, J.: Sensor placement with a telescoping compliant mechanism, In: 4th European Conference of the International Federation for Medical and Biological Engineering, Antwerpen: 2008.11.23–27, Berlin, Springer, 1987–1989, doi:10.1007/978-3-540-89208-3_473, 2009

[39] Hale, L. C.: Principles and Techniques for Designing Precision Machines, Doctoral thesis, Cambridge, Massachusetts Institute of Technology, 1999

[40] Hartmann, L., Feierabend, M., Zentner, L.: Über die Wirkung einer mechanischen Vorspannung auf die Deformation eines asymmetrischen Federbügel-Mechanismus, In: 10. Kolloquium Getriebetechnik, 2013.09.11–13, Ilmenau, Univ.-Verl., 373–390, ISSN 2194-9476, 2013

[41] Hartmann, L., Uhlig, R., Zentner, L.: Analytische und messtechnische Untersuchungen zum Schwingungsverhalten von Zungenventilen in Membranverdichtern, Ilmenau, Univ.-Verl., Berichte der Ilmenauer Mechanismentechnik 1, ISSN 2194-9476, 45–54, 2012

[42] Henein, S.: Conception des guidages flexibles, Lausanne, Presses polytechniques et universitaires romandes, ISBN 978-2-88074-481-4, 2001

[43] Henein, S., Spanoudakis, P., Droz, S., Myklebust, L. I., Onillon, E.: Flexure pivot for aerospace mechanisms, In: Proceedings of the 10th European Space Mechanisms and Tribology Symposium, San Sebastian, Spain, 2003

[44] Henning, S., Linß, S., Zentner, L.: detasFLEX – A computational design tool for the analysis of various notch flexure hinges based on non-linear modeling, Mechanical Sciences, 9, 389–404, doi:10.5194/ms-9-389-2018, 2018

[45] Hopkins, J. B., Culpepper, M. L.: Synthesis of multi-degree of freedom, parallel flexure system concepts via freedom and constraint topology (FACT) – Part I: Principles, Precision Engineering, 34, 259–270, doi:10.1016/j.precisioneng.2009.06.008, 2010

[46] Horie, M., Hoshikawa, Y., Kamiya, D.: Optimum Design of Mass Distribution of the Injection Molding Pantograph Mechanism with Constant Output Link Orientation, In: Proceedings of the 1st Workshop on Microactuators and Micromechanisms (MAMM), Aachen, Germany, 31–39, doi:10.1007/978-94-007-2721-2_4, 2010

[47] Howell, L.: Compliant mechanisms. New York, Wiley, ISBN 9780471384786, 2001

[48] Howell, L. L., Midha, A.: A Method for the Design of Compliant Mechanisms with Small-Length Flexural Pivots, Journal of Mechanical Design, 116, 280–290, doi:10.1115/1.2919359, 1994

[49] Howell, L. L., Magleby, S. P., Olsen, B. M.: Handbook of Compliant Mechanisms, Chichester, Wiley, ISBN 0-471-38478-X, 2013

[50] Issa, M., Zentner, L.: Sensorelemente aus leitfähigem Silikon für einen nachgiebigen Greifer, Ilmenau, Univ.-Verl., Berichte der Ilmenauer Mechanismentechnik 1, 65–78, ISSN 2194-9476, 2012

[51] Issa, M., Petkovic, D., Pavlovic, N.D., Zentner, L.: Sensor elements made of conductive silicone rubber for passively compliant gripper, The International Journal of Advanced Manufacturing Technology, 69, 1527–1536, doi:10.1007/s00170-013-5085-8, 2013

[52] Jensen, B. D., Howell, L. L.: The modeling of cross-axis flexural pivots, Mechanism and Machine Theory, 37, 461–476, doi:10.1016/S0094-114X(02)00007-1, 2002

[53] Kaletsch, T.: Systematische Untersuchung von monolithischen Mehrfachdrehgelenken und der Anwendung in nachgiebigen Mechanismen, Master‘s thesis, Ilmenau, Technische Universität Ilmenau, 2017

[54] Kamke, E.: Differentialgleichungen Lösungsmethoden und Lösungen, I – gewöhnliche Differentialgleichungen, Leipzig, Geest & Portig, 1959

[55] Li, L., Liu, H., Zhao, H.: A Compact 2-DOF Piezoelectric-Driven Platform Based on "Z-Shaped" Flexure Hinges, Micromachines, 8, 245, doi:10.3390/mi8080245, 2017

[56] Li, Q., Pan, C., Xu, X.: Closed-form compliance equations for power-function-shaped flexure hinge based on unit-load method, Precision Engineering, 37, 135–145, doi:10.1016/j.precisioneng.2012.07.010, 2013

[57] Lin, R., Zhang, X., Long, X., Fatikow, S.: Hybrid flexure hinges, Review of Scientific Instruments, 84, 085004-1-14, doi:10.1063/1.4818522, 2013

[58] Linß, S.: Ein Beitrag zur geometrischen Gestaltung und Optimierung prismatischer Festkörpergelenke in nachgiebigen Koppelmechanismen, Doctoral thesis, Ilmenau, Technische Universität Ilmenau, urn:nbn:de:gbv:ilm1-2015000283, 2015

[59] Linß, S., Erbe, T., Theska, R., Zentner, L.: The influence of asymmetric flexure hinges on the axis of rotation, In: Proceedings of the 56th International Scientific Colloquium, Ilmenau, Germany, urn:nbn:de:gbv:ilm1-2011iwk-006:6, 2011

[60] Linß, S., Erbe, T., Zentner, L.: On polynomial flexure hinges for increased deflection and an approach for simplified manufacturing, In: Proceedings of the 13th World Congress in Mechanism and Machine Science, Guanajuato, Mexico, A11_512, 2011

[61] Linß, S., Gräser, P., Räder, T., Henning, S., Theska, R., Zentner, L.: Influence of geometric scaling on the elasto-kinematic properties of flexure hinges and compliant mechanisms, Mechanism and Machine Theorie, 125, 220–239, doi:10.1016/j.mechmachtheory.2018.03.008, 2018

[62] Linß, S., Milojevic, A., Pavlovic, N. D., Zentner, L.: Synthesis of Compliant Mechanisms based on Goal-Oriented Design Guidelines for Prismatic Flexure Hinges with Polynomial Contours, In: Proceedings of the 14th World Congress in Mechanism and Machine Science, Taipei, Taiwan, doi:10.6567/IFToMM.14TH.WC.PS10.008, 2015

[63] Linß, S., Milojevic, A., Zentner, L.: On the Influence of Flexure Hinge Geometry on the Motion Range and Precision of Compliant Gripping Mechanisms, In: Proceedings of the 2nd International Conference Mechanical Engineering in XXI Century, Niš, Serbia, 255–260, 2013

[64] Linß, S., Milojevic, A.: Model-based design of flexure hinges for rectilinear guiding with compliant mechanisms in precision systems, Ilmenau, Univ.-Verl., Berichte der Ilmenauer Mechanismentechnik 1, ISSN 2194-9476, 13–24, 2012

[65] Linß, S., Schorr, P., Henning, S.; Zentner, L.: Contour-independent design equations for the calculation of the rotational properties of commonly used and polynomial flexure hinges, In: Proceedings of the 59th Ilmenau Scientific Colloquium, Ilmenau, Germany, urn:nbn:de:gbv:ilm1-2017iwk-001:5, 2017

[66] Linß, S., Schorr, P., Zentner, L.: General design equations for the rotational stiffness, maximal angular deflection and rotational precision of various notch flexure hinges, Mechanical Sciences, 8, 29–49, doi:10.5194/ms-8-29-2017, 2017

[67] Liu, M., Zhang, X., Fatikow, S.: Design and analysis of a multi-notched flexure hinge for compliant mechanisms, Precision Engineering, 48, 292–304, doi:10.1016/j.precisioneng.2016.12.012, 2017

[68] Lobontiu, N.: Compliant Mechanisms: Design of Flexure Hinges, Boca Raton, Fla., CRC Press, ISBN 0-8493-1367-8, 2003

[69] Lobontiu, N., Cullin, M.: In-plane elastic response of two-segment circular-axis symmetric notch flexure hinges: The right circular design, Precision Engineering, 37, 542–555, doi:10.1016/j.precisioneng.2012.12.007, 2013

[70] Lobontiu, N., Paine, J.: Parabolic and hyperbolic flexure hinges, Precision Engineering, 26, 183–192, doi:10.1016/S0141-6359(01)00108-8, 2002

[71] Lobontiu, N., Paine, J. S. N., Garcia, E., Goldfarb, M.: Corner-Filleted Flexure Hinges, Journal of Mechanical Design, 123, 346–352, doi:10.1115/1.1372190, 2001

[72] Lotti, F., Vassura, G.: A novel approach to mechanical design of articulated fingers for robotic hands, In: Proceedings of IEEE/RSJ International Conference on Intelligent Robots and Systems IROS, Lausanne, Switzerland, 1687–1692, doi:10.1109/IRDS.2002.1043998, 2002

[73] Lotze, W.: Die ebene Kinematik von Biegefederaufhängungen, Doctoral thesis, Dresden, Technische Universität Dresden, 1965

[74] Love, A. E.: Treatise on the Mathematical Theory of Elasticity, 4th Edition, Published by Dover Pubns, 1927

[75] Martin, J., Robert, M.: Novel Flexible Pivot with Large Angular Range and Small Center Shift to be Integrated into a Bio-Inspired Robotic Hand, Journal of intelligent material systems and structures, 22, 1431–1438, doi:10.1177/1045389X11412639, 2011

[76] Meng, Q.: A Design Method for Flexure-Based Compliant Mechanisms on the Basis of Stiffness and Stress Characteristics, Doctoral thesis, Bologna, University of Bologna, 2012

[77] Meng, Q., Berselli, G., Vertechy, R., Castelli, V. P.: An Improved Method for Designing Flexure-Based Nonlinear Springs, In: Proceedings of the 36th Mechanisms and Robotics Conference (ASME 2012), Chicago, USA, 211–219, doi:10.1115/DETC2012-70367, 2012

[78] Meng, Q., Li, Y., Xu, J.: New empirical stiffness equations for corner-filleted flexure hinges, Mechanical Sciences, 4, 345–356, doi:10.5194/ms-4-345-2013, 2013

[79] Merriam, E. G., Howell, L. L.: Lattice flexures: Geometries for stiffness reduction of blade flexures, Precision Engineering, 45, 160–167, doi:10.1016/j.precisioneng.2016.02.007, 2016

[80] Müller, H. H., Magnus, K.: Übungen zur technischen Mechanik, Stuttgart, Verlag Teubner, p. 128, ISBN 3-519-22325-2, 1988

[81] Niu, Y., Song, Z., Dai, J. S.: Design of the Planar Compliant Five-bar Mechanism for Self-aligning Knee Exoskeleton, In: Proceedings 4th International conference on Reconfigurable Mechanisms and Robots (ReMAR), Delft, The Netherlands, doi:10.1109/REMAR.2018.8449858, 2018

[82] Noll, T.: Three axis rotational flexure joints of high axial stiffness, Precision Engineering, 26, 460–465, doi:10.1016/S0141-6359(02)00148-4, 2002

[83] Paros, J. M., Weisbord, L.: How to design flexure hinges, Machine design, 25, 151–156, 1965

[84] Pavlovic, N. D., Petkovic, D., Pavlovic, N. T.: Optimal selection of the compliant mechanism synthesis method, In: The International Conference Mechanical Engineering in XXI Century, Niš, Serbia, 247–250, 2010

[85] Pavlovic, N. T., Pavlovic, N. D.: Mobility of the compliant joints and compliant mechanisms, Theoretical Applied Mechanics, 32, 341–357, 2005

[86] Pei, X., Yu, J., Zong, G., Bi, S., Yu, Z.: Analysis of Rotational Precision for an Isosceles-Trapezoidal Flexural Pivot, Journal of Mechanical Design, 130, 052302-1-9, doi:10.1115/1.2885507, 2008

[87] Raatz, A.: Stoffschlüssige Gelenke aus pseudo-elastischen Formgedächtnislegierungen in Parerellrobotern, Doctoral thesis, Essen, Vulkan-Verl, ISBN 3–8027–8691–2, 2006

[88] Ragulskis, K. M., Arutunian, M. G., Kochikian, A. V., and Pogosian, M. Z.: A Study of Fillet Type Flexure Hinges and their Optimal Design, Vibration Engineering, 3, 447–452, 1989

[89] Risto, U.: Zur Charakterisierung und Anwendung des Durschlagverhaltens von nachgiebigen rotationssymmetrischen Strukturen, Doctoral thesis, Ilmenau, Univ.-Verl., BIMT, urn:nbn:de:gbv:ilm1-2013000467, 2013

[90] Risto, U., Zentner, L., Uhlig, R.: Elastic structures with snap-through characteristic for closing devices, In: Proceedings of the 53th International Scientific Colloquium, Ilmenau, Germany, 8–12, 2008

[91] Rosengren, L.: Universal joint, European Patent Application, EP0568514A3, filed 26. Apr. 1993

[92] Schotborgh, W. O., Kokkeler, F. G., Tragter, H., Houten, F. J. A. M. van: Dimensionless design graphs for flexure elements and a comparison between three flexure elements, Precision Engineering, 29, 41–47, doi:10.1016/j.precisioneng.2004.04.003, 2005

[93] Smith, S. T., Badami, V. G., Dale, J. S., Xu, Y.: Elliptical flexure hinges, Review of Scientific Instruments, 68, 1474–1483, doi:10.1063/1.1147635, 1997

[94] Tanik, E., Parlaktas, V.: A new type of compliant spatial four-bar (RSSR) mechanism, Mechanism and Machine Theorie, 46, 593–607, doi:10.1016/j.mechmachtheory.2011.01.004, 2011

[95] Thorp, A. G.: Flexure pivots-design formulas and charts, Product Engineering, 24, 192–200, 1953

[96] Tian, Y., Shirinzadeh, B., Zhang, D., Zhong, Y.: Three flexure hinges for compliant mechanism designs based on dimensionless graph analysis, Precision Engineering, 34, 92–101, doi:10.1016/j.precisioneng.2009.03.004, 2010

[97] Trease, B. P., Moon, Y.-M., Kota, S.: Design of Large-Displacement Compliant Joints, Journal of Mechanical Design, 127, 788–798, doi:10.1115/1.1900149, 2005

[98] Tseytlin, Y. M.: Notch flexure hinges: An effective theory, Review of Scientific Instruments, 73, 3363–3368, doi:10.1063/1.1499761, 2002.

[99] Vallance, R. R., Haghighian, B., Marsh, E. R.: A unified geometric model for designing elastic pivots, Precision Engineering, 32, 278–288, doi:10.1016/j.precisioneng.2007.10.001, 2008

[100] VDI/VDE-Richtlinie, VDI/VDE 2252: Feinwerkelemente, Führungen, Federgelenke, Berlin, Beuth-Verlag, 1990

[101] VDI-Richtlinie, VDI 2740-2: Mechanische Einrichtungen in der Automatisierungstechnik – Führungsgetriebe, Berlin, Beuth-Verlag, 2002

[102] Venanzi, S., Giesen, P., Parenti-Castelli, V.: A novel technique for position analysis of planar compliant mechanisms, Mechanism and Machine Theory, 40, 1224–1239, doi:10.1016/j.mechmachtheory.2005.01.009, 2005

[103] Wang, R., Zhou, X., Zhu, Z.: Development of a novel sort of exponent-sine-shaped flexure hinges, Review of Scientific Instruments, 84, 095008-1-10, doi:10.1063/1.4821940, 2013

[104] Winder, B. G., Magleby, S. P., Howell, L. L.: A Study of Joints Suitable for Lamina Emergent Mechanisms, In: Proceedings of the ASME 2008 International Design Engineering Technical Conferences and Computers and Information in Engineering Conference (IDETC/CIE2008), New York, USA, 339–349, doi:10.1115/DETC2008-49914, 2008

[105] Wittwer, J. W., Howell, L. L.: Mitigating the Effects of Local Flexibility at the Built-In Ends of Cantilever Beams, Journal of Applied Mechanics, 71, 748–751, doi:10.1115/1.1782913, 2004

[106] Wuest, W.: Blattfedergelenke für Meßgeräte, Feinwerktechnik FWT; Zeitschrift für Feinmechanik, Optik, Elektronik, Meß- und Prüfwesen, Entwicklung, Fertigung, 54, 167–170, 1950

[107] Xu, Q.: Design and Implementation of Large-Range Compliant Micropostioning Systems, Singapore, John Wiley & Sons, ISBN 9781119131441, 2016

[108] Yong, Y. K., Lu, T.-F., Handley, D. C.: Review of circular flexure hinge design equations and derivation of empirical formulations, Precision Engineering, 32, 63–70, doi:10.1016/j.precisioneng.2007.05.002, 2008

[109] Yu, A., Xi, F., Li, B.: A Shape-Adaptive Mechanism with SMP Jointed Panels – Preiliminary Study, In: Proceedings 4th International conference on Reconfigurable Mechanisms and Robots (ReMAR), Delft, The Netherlands, doi:10.1109/REMAR.2018.8449862, 2018

[110] Yu, J., Li, S., Pei, X., Bi, S., Zong, G.: A unified approach to type synthesis of both rigid and flexure parallel mechanisms, Science in China, 54, 1206–1220, doi:10.1007/s11431-011-4324-1, 2011

[111] Yu, J., Zong, G., Yu, Z., Bi, S.: A New Family of Large-Displacement Flexural Pivots, In: Proceedings of ASME 2007 International Design Engineering Technical Conferences & Computers and Information in Engineering Conference (IDETC/CIE 2007), Las Vegas, USA, 93 – 100, doi:10.1115/DETC2007-34977, 2007

[112] Zeidler, E., Hackbusch, W., Schwarz, H.: Teubner-Taschenbuch Mathematik, Zeidler, E. (Ed.), Leipzig, Teubner Verlagsgesellschaft, ISBN 3-8154-2001-6, 1996

[113] Zelenika, S., Munteanu, M. G., Bona, F. De: Optimized flexural hinge shapes for microsystems and high-precision applications, Mechanism and Machine Theory, 44, 1826–1839, doi:10.1016/j.mechmachtheory.2009.03.007, 2009

[114] Zentner, L.: Nachgiebige Mechanismen, München, De Gruyter Oldenbourg, ISBN 978-3-486-76881-7, 2014

[115] Zentner, L.: Untersuchung und Entwicklung nachgiebiger Strukturen basierend auf innendruckbelasteten Röhren mit stoffschlüssigen Gelenken, Ilmenau, ISLE Verlag, ISBN 3-932633-77-6, 2003

[116] Zentner, L.: Mathematischer Formalismus zur Bildung von Starrkörpermodellen für nachgiebige Mechanismen, In: 9. Kolloquium Getriebetechnik, Berger, M. (Ed.), Chemnitz, Univ.-Verl., 227–244, ISBN 978-3-941003-40-8, 2011

[117] Zentner, L.: Klassifikation nachgiebiger Mechanismen und Aktuatoren, Ilmenau, Univ.-Verl., Berichte der Ilmenauer Mechanismentechnik 1, 3–12, ISSN 2194-9476, 2012

[118] Zentner, L.: Modelling and Application of Hydraulic Spider Leg Mechanism, Spider Ecophysiology, Nentwig, W. (Ed.), Berlin, Heidelberg, Springer-Verlag, ISBN 978-3-642-33988-2, 2013

[119] Zentner, L., Böhm, V.: Zum Verformungsverhalten nachgiebiger Mechanismen, Konstruktion Zeitschrift für Produktentwicklung und Ingenieur-Werkstoffe, 60, 67–71/74, 2008

[120] Zentner, L., Böhm, V.: On the Mechanical Compliance of Technical Systems, Mechanical Engineering, Gokcek, M. (Ed.), InTech, doi:10.5772/26379, 2012

[121] Zentner, L., Turkevi-Nogy, N., Böhm, V.: Entwicklung von fluidischen Elastomeraktoren, In: Proceedings of the 47th International Scientific Colloquium, Ilmenau, Germany, 2002

[122] Zentner, L., Griebel, S., Wystup, C., Hügl, S., Rau, T. S., Majdani, O.: Synthesis process of a compliant fluidmechanical actuator for use as an adaptive electrode carrier for cochlear implants, Mechanism and Machine Theory, 112, 155-171, doi:10.1016/j.mechmachtheory.2017.02.001, 2017

[123] Zettl, B., Szyszkowski, W., Zhang, W. J.: On Systematic Errors of Two-Dimensional Finite Element Modeling of Right Circular Planar Flexure Hinges, Journal of Mechanical Design, 127, 782–787, doi:10.1115/1.1898341, 2005

[124] Zhu, B. L., Zhang, X. M., Fatikow, S.: Design of single-axis flexure hinges using continuum topology optimization method, Science in China, 57, 560–567, doi:10.1007/s11431-013-5446-4, 2014

[125] Zinn, M., Khatib, O., Roth, B., Salisbury, J. K.: A new actuation approach for human friendly robot design, The International Journal of Robotics Research, 23, No.4-5, 379–398, doi:10.1177/0278364904042193, 2004

[126] Светлицкий, В.А.: Механика стержней, Москва, Высшая школа, 1987

Index

https://doi.org/10.1515/9783110479744-175

Bei Fragen zur Produktsicherheit wenden Sie sich bitte an:
If you have any questions regarding product safety,
please contact:

Walter de Gruyter GmbH
Genthiner Straße 13
10785 Berlin
productsafety@degruyterbrill.com